Praise for *Obaysch: A Hippopotamus in Victorian London*

'From its very first word this book drew me in, made me cry and elicited understanding and unease … This thoughtful, meticulously researched book begs to be read by animal studies scholars and anyone concerned about the plight of species other than ourselves.'
—Carol Freeman, University of Tasmania, author of *Paper Tiger: How Pictures Shaped the Thylacine*

'John Simons is a skilful storyteller and *Obaysch* is a compelling read. Meticulously researched and generously illustrated, the book fulfils Simons' determination "to treat Obaysch as an actor in his own life" at the same time as exploring how this unfortunate hippo became "the most important animal of the Victorian era". The result is fine addition to the Animal Publics series, and a significant contribution to the emerging field of animal biography.'
—Steve Baker, Emeritus Professor of Art History, University of Central Lancashire

'John Simons' richly exhaustive account of nineteenth-century hippomania engages with imperialism, Orientalism, progress, and the cultural history of Europe … Poignant and empathetic, this account of an animal's appropriation and exploitation is one of those books that unfurls more about its moment in time than you could have imagined when you picked it up.'
—Randy Malamud, Regents' Professor of English, Georgia State University, Atlanta, author of *Reading Zoos: Representations of Animals and Captivity*

'Simons' breadth of reference, his often witty commentary, and even his footnotes (what connection can there possibly be between the notorious drug lord Pablo Escobar and hippopotamuses?) make fascinating reading.'
—Helen Tiffin, University of Wollongong, author of *Wild Man from Borneo: A Cultural History of the Orangutan*

ANIMAL PUBLICS

Melissa Boyde & Fiona Probyn-Rapsey, Series Editors

The Animal Publics series publishes new interdisciplinary research in animal studies. Taking inspiration from the varied and changing ways that humans and non-human animals interact, it investigates how animal life becomes public: attended to, listened to, made visible, included, and transformed.

Animal Death
Ed. Jay Johnston and Fiona Probyn-Rapsey

Animal Welfare in Australia: Politics and Policy
Peter Chen

Animals in the Anthropocene: Critical Perspectives on Non-human Futures
Ed. The Human Animal Research Network Editorial Collective

Cane Toads: A Tale of Sugar, Politics and Flawed Science
Nigel Turvey

Engaging with Animals: Interpretations of a Shared Existence
Ed. Georgette Leah Burns and Mandy Patersonn

Fighting Nature: Travelling Menageries, Animal Acts and War Shows
Peta Tait

The Flight of Birds: A Novel in Twelve Stories
Joshua Lobb

A Life for Animals
Christine Townend

Obaysch: A Hippopotamus in Victorian London
John Simons

Obaysch

A Hippopotamus in Victorian London

John Simons

SYDNEY UNIVERSITY PRESS

First published by Sydney University Press
© John Simons 2019
© Sydney University Press 2019

Reproduction and communication for other purposes
Except as permitted under the Act, no part of this edition may be reproduced, stored in a retrieval system, or communicated in any form or by any means without prior written permission. All requests for reproduction or communication should be made to Sydney University Press at the address below:

Sydney University Press
Fisher Library F03
University of Sydney NSW 2006
AUSTRALIA
sup.info@sydney.edu.au
sydney.edu.au/sup

 A catalogue record for this book is available from the National Library of Australia.

ISBN 9781743325865 paperback
ISBN 9781743325896 epub
ISBN 9781743323267 mobi

Cover image: *Obaysch, the Hippopotamus, London Zoo* (1852), Don Juan, Comte de Montizón (1822–87) (photographer), salt print laid on card, RCIN 2905523. Probably acquired by Queen Victoria and Prince Albert in 1854, following the Photographic Society Exhibition, London. Royal Collection Trust © Her Majesty Queen Elizabeth II 2017.

Cover design by Miguel Yamin.

Contents

List of Plates	vii
Preface and Acknowledgements	xi
Prologue	xvii
Introduction: Why Obaysch?	1
1 The Life and Times of Obaysch the Hippopotamus	33
2 The Several Meanings of Hippos	133
3 A Bloat of Other European Hippos	171
Postscript: The Unhappy Hippopotamus	205
Obaysch and His Bloat	209
A Note on Sources	211
Bibliography	213
Index	221

List of Plates

Plate 1 William Blake's Behemoth from his illustrations to the *Book of Job* (1826).

Plate 2 Late-eighteenth-century engraving of a hippopotamus.

Plate 3 'Hippopotamus' engraving from the painting by Daniell published by Cadell and Davies (London, 1807).

Plate 4 'The Haunt of Behemoth', *Illustrated Sporting and Dramatic News*, 21 November 1874.

Plate 5 Obaysch and Hamet. 'The Hippopotamus in the Gardens of the Zoological Society, Regent's Park', *Illustrated London News*, 1 June 1850.

Plates 6 and 7 Watercolours of Obaysch by Joseph Wolf (1850), based on sketches sent from Egypt by Charles Murray.

Plate 8 'The Hippopotamus in the Gardens of the Zoological Society, Regent's Park', *Illustrated London News*, 8 June 1850.

Plate 9 'The Hippopotamus in His New Bath in the Zoological Gardens, Regent's Park', *Illustrated London News*, 14 June 1851.

Plate 10 Obaysch meets Adhela: 'The Female Hippopotamus at the Zoological Gardens, Regent's Park', *Illustrated London News*, 12 August 1854.

Obaysch

Plate 11 'The Baby Hippopotamus at the Zoo', *The Graphic*, 1871.

Plate 12 'The Hippopotamus and Her Young One at the Zoological Gardens', *The Graphic*, 1872.

Plate 13 'Sunday Afternoon at the Zoological Gardens – Beauty and the Beast', *The Graphic*, 21 November 1891.

Plate 14 'The Hippopotami in Their New Tank at Central Park', *Harper's Weekly*, 29 September 1888.

Plate 15 'Mr Gordon Cumming's South Africa Entertainment – view of the Limpopo with a Herd of Hippopotami', *Illustrated London News*, 5 January 1856.

Plate 16 'Hippopotamus and Salee', *The Graphic*, 26 January 1884.

Plate 17 'Hippopotamus Hunting in Angola, West Africa', *Illustrated London News*, 17 January 1880.

Plate 18 'Native Hunters Harpooning a Hippopotamus'. *Illustrated Sporting and Dramatic News*, 21 November 1874.

Plate 19 'Hippopotamus Shooting', *Illustrated Sporting and Dramatic News*, 21 November 1874.

Plate 20 'Native Hunters Hauling a Hippopotamus Ashore.' *Illustrated Sporting and Dramatic News*, 21 November 1874.

Plate 21 'The Adventure with a Hippopotamus', source unknown.

Plate 22 'Canoe destroyed by a Hippopotamus on the River Zambesi, South Africa', *Illustrated London News*, 19 May 1866.

Plate 23 'Boat Capsized by a Hippopotamus Robbed of Her Young', *Illustrated London News*, 7 November 1857.

Plate 24 'On the Victoria Nyanza: a Hippopotamus Attacks a Shooting Party', *Illustrated London News*, 29 August 1899.

Plate 25 A sketch by Robert Baden-Powell of himself shooting hippos, from his *Lessons from the Varsity of Life* (London: C. Arthur Pearson Ltd, 1933).

List of Plates

Plate 26 'The New Comer – a Sketch in the Depôt of an Importer of Animals', *Harper's Weekly*, 29 September 1888.

Plate 27 'Young Hippopotamus (*liberiensis*) recently landed at Liverpool (now dead)', *The Graphic*, 29 March 1873.

Plate 28 'The Home of the Hippopotamus', *Illustrated London News*, 11 February 1899.

Preface and Acknowledgements

This is the third of my books to deal with Obaysch. The first, *Rossetti's Wombat*, looked at him simply as an example of an exotic animal in the zoo alongside many others and, especially, the curious collection which wandered the dark corridors and leafy garden of Dante Gabriel Rossetti's mansion in Chelsea. The second, *The Tiger that Swallowed the Boy*, placed him in the more general context of the flow of animals around the British Empire and the international wild animal trade in the nineteenth century as well as sketching out his specific impact on the success of London Zoo. But, after I had thought about him for about ten years I came to the belief that neither of these works dealt with him in the detail he deserves.

From the point of view of the continued viability of the Zoological Society of London's Regent's Park Zoological Gardens and its establishment as one of the main destinations on any trip to London, from the point of view of zoology and of stimulating popular interest in science, from the point of view of the establishment of routes for animals to travel around the Empire, from the point of view of the big picture politics of the British and French expansion into Africa and the waning influence of the

Ottoman Empire, Obaysch is, quite simply, the most important animal of the Victorian era.

This is much the same as saying that he was the most important animal of the nineteenth century.

Such an animal deserves and repays proper study.

And, of course, this work sits among others which have also considered him. There are four chief differences between my work and all the others. The first is that this work (like all my writings on the history of animals) is designed to promote the idea that animals have agency and are, therefore, proper subjects of our moral concern. Putting it bluntly, scholarship about animals ought to encourage people not only to think about animals more carefully but also to treat them more kindly. For me, Obaysch is not just a term in the semiotic chain or a political or cultural symbol. He is presented as a real living being who had thoughts and feelings (although we cannot know what these were) that call for and deserve acknowledgement. He was also the victim of a great wrong when he was captured in the wild and transported to live a highly restricted and unnatural life in a cold northern city. This undoubtedly harmed him and this also calls for acknowledgement. The second difference is that I have deployed a wider range of Obaysch-related sources than ever before. In fact there was much more material to be found than I could use. It would, for example, have been possible to have a full chapter simply on verse, cartoons and images featuring hippos, in addition to those verses and those cartoons mentioned in the text.

I have subjected those sources I have used to some critical analysis, in particular to address the reliability or otherwise of traditions about Obaysch, which are often repeated from one secondary source to another as if they were undoubted facts. This has led me to cast some doubts on a number of details that are frequently used to support the idea of a craze of 'hippomania'. While there is no doubt that Obaysch was astonishingly popular there is, nonetheless, reason to think that the idea of hippomania as a freestanding craze separate from other nearly contemporary scientific crazes – rather than part of the general Victorian enthusiasm for science and natural history – has yet to be fully established as an historical entity. Thirdly, my work looks at Obaysch in the context of an international trade in hippos in the second half of the nineteenth century and also, importantly I think, in the context of his position as the alpha male of a small bloat (my preference out of the several names for a social group of hippos) living in the Zoological Gardens. By studying his mate Adhela and Guy Fawkes, the only one of his three offspring who survived into adulthood, I have, I believe, been able to show how Obaysch's image was constructed and manipulated by his various keepers and by the Victorian news media. Finally, this work acknowledges that Obaysch's life started before he first appeared in the Zoological Society's network and that his one year – maybe only a few months – as a wild hippopotamus is a key to understanding much else about him.

In reminding the reader of this, I am also attempting to defuse the tendency in animal studies by which animals often seem only to present themselves to us when they are part of a chain of human perception and activity. It seems to me that scholarship can treat animals in a neo-Berkeleyan sense and leaves them as mere effects of being until they are perceived or acted on in some way (usually a bad way) by humans ('*esse est percipi*', as George Berkeley wrote: to be is to be perceived). I don't think this is an intentional thing but reading thousands of pages of books about animals over many years has convinced me that many people who study animals don't really believe they exist. Or if they do exist they exist solely to make some contingent political point.

I have speculated elsewhere on the enterprise of writing the biography of a dead animal. In April 2017 Patrick, a rescue wombat who had been well cared for in the Ballarat Wildlife Park for thirty-one years, passed away having comfortably broken every record for confirmed wombat longevity. I met Patrick in early 2006 and had the privilege of going into his burrow when I was writing the biography of Top, a wombat who lived less than a few months in Victorian England, and I have kept in touch with him through his Facebook page more or less ever since. This sounds whimsical of course but the fact is that thousands of people around the world came to know Patrick and recognised his agency. This is not just trivial anthropomorphism although it superficially appears to be. Patrick lived a highly social life as part of a media-based global community made possible by digital technology and, as such, he acquired a variety of personhood. When I compare Obaysch with Patrick I can identify some similarities as Obaysch also lived as the centre of a social network. This was not, of course, interactive, but it was as genuinely global as anything in the mid nineteenth century could be: people in Auckland or Bendigo were as likely to be reading about Obaysch as people in London or Edinburgh. After reading the many accounts of seeing Obaysch that have come down to us, there can be no doubt that he was recognised by many who encountered him as having the same kind of claim to personhood as Patrick. This work is not an exploration of that personhood, if indeed it existed, but it does treat it as a way of seeing Obaysch which goes beyond the merely descriptive or constructionist.

This book was finished soon after I had retired from a long career spanning some forty-one years as a student, teacher and senior manager in English, American and Australian universities and came to live on the beautiful island of Tasmania, where I can see whales swimming past, weave and dye cloth, make jam, churn butter and not think very much about government education policy. I was immensely lucky to have enjoyed pretty much every day of my career: to spend one's days talking with interesting colleagues, seeing fabulous young people grow up and go out into the world and older people enjoy the fulfillment of their interests that they missed in their younger days, and to get paid to pursue one's interests in the back streets of the history of animals, is surely a bundle of the greatest privileges that anyone can be granted. But the last few years were the best, and I'd like to acknowledge Professor Bruce

Dowton, Vice Chancellor of Macquarie University, with whom I was privileged to work on the university's executive as Deputy Vice Chancellor (Academic) for a number of wonderful years. I am grateful to Bruce for many things: not least his recognition that someone who finds life generally amusing can still do a grown-up's job reasonably well. Not all vice chancellors are like that. I wish him and Macquarie University every success.

I'd like to thank Melissa Boyde of the University of Wollongong, who commissioned this work and has always been kind enough to involve me in her various animal-related projects. I'd also like to thank Agata Mrva-Montoya and Denise O'Dea, my editorial team at Sydney University Press: their careful attention and wise suggestions have been a constant help and source of improvement.

I usually note at the end of my prefaces the feast day on which the main work on the book was finished. Readers will see a happy serendipity in this case.

But, as always, my greatest debt is to my wife Kate, who helped me see that there really was a full length study in Obaysch. As I write I can hear the rhythm of her saw as she works on a piece of silver jewellery which will, no doubt, turn out exquisitely like everything else to which she turns her creative spirit.

John Simons
Taroona
The Feast of Saint Hippolytus,
Martyr of Rome, 2017

Prologue

She had been far far too far north when the cooling started. She lived right on the edge of the hippopotamus range. But up there the feeding was still plentiful as more and more of the animals drifted south. The hunters had also gone, following them. But now the grasses weren't growing back again. The river was too cold for her. She was hungry and she couldn't remember when she had last seen another hippo. No mate, no children any more. A hippo without a bloat and that was a bad thing. All the others had gone to feed the hunters or had joined the weary trek south, a hard thing for a hippo as it meant not being able to spend the days in water. It had seemed to her wise to stay and enjoy the lack of competition for the increasingly sparse grasses and weeds. She didn't know what was down there except more competition for food, which was getting scarce, and more hunters. But now she stood alone, looking back to the grey and white skies over what had been her home. When she was born thirty years ago they had still been blue.

She had been walking for a month through a dying landscape stripped of all the things she needed to keep her going. She was hungry and she was weak and she was tired. She lay down, not to sleep but just to rest, but even so she drifted off. When she woke up

Obaysch

she found her legs wouldn't lift her any more. She struggled to rise but her front legs, which had served her so well for so many years and so many miles, simply buckled at every attempt and after six or seven tries she lay there panting and only half awake. Then the first arrow hit her and then another. She lay there wondering why she had not smelled out the hunters as they approached. Now she heard them closing in and she did smell the sharp tang of the bear and wolf skins they wrapped themselves with. One placed a foot on her shoulder and thrust hard with his spear down into her body. The sharp tip slid down to her heart.

She was the last hippopotamus of the inter-glacial period in the British Isles.

He was still young when he was dragged from the river. He'd been separated from the other hippos and was grazing on his own, enjoying the sweetness of the Nile cabbages and the warmth of the water when the net descended. Before he understood what was happening a strong rope had been looped round his front legs and pulled tighter and tighter as he was dragged from the river. He grunted in pain. A cage awaited him and when he had been pulled in, struggling, the rope was cut. A handful of grass was thrust in but he was too enraged and frightened to eat it. He pushed his head against the iron bars but they were too strong for him and eventually he lay exhausted and breathing hard in quick wheezy gasps. The cage was pushed inch by inch up a ramp onto a large cart and when it was secured a team of twenty oxen strained and he was moving.

After a day his cart joined many others, hundreds of them, with hundreds of animals. Lions, leopards, ostriches, rhinoceros, zebras, quaggas, all confined and roaring or prowling or simply sitting dazed. There were nearly ten thousand all together and he saw at least two more hippos. The caravan set off and a great herd of elephants walked in chains behind it. For three weeks it travelled

Prologue

slowly across a dry and parched land. He was hungry and thirsty and was burned by the sun. The other hippos died after a couple of weeks. Then they reached the sea and the animals, still in their cages or chains, were loaded onto a fleet of ships. He was on the sea for about a week and then his ship nosed down a wide brown river. His cage was unloaded and he was put on another cart and dragged through crowded streets. People stared at him as if they had never seen his like before. Finally, they came to a ramp and his cage was taken underground. The gate was opened and he staggered out into a small pen. It had little light but there was a big pile of a strange tasting grass and a large bowl of water. He ate and drank greedily. For a month he lived in the half-light, the sun filtering only occasionally through a grate above his pen, but he was well fed on the strange grass and given plenty of water. At the end of that month he heard shouting and screams that went on half the day. He could smell the iron-sweet tang of blood and hear the angry and anxious roars of lions and the groans and cheers of a large crowd. Then the gate at the other end of his pen was opened and he was prodded to trot down a narrow corridor and up a ramp. There another gate was opened and he blinked in the bright sunshine that streamed in and blinded him. When his eyes accustomed themselves to the glare he saw the arena, sandy and soaked in blood. In front of him were two men, one with a sword and one with a spear. They gestured and shouted at him to come. He tried to turn back to the darkness of his pen and felt thirsty, as if the light had sucked all the moisture from him. But sharp stabs from pointed sticks in his rump drove him forward snorting in pain and the gate was slammed shut behind him. The men gestured and capered and the crowd cheered and laughed. He felt rage build within him and trotted forward half blind with anger. The man with the sword started to run away and as he chased him he felt the spear pierce his side and then as he staggered a sharp knife slicing though the muscles and tendons of his back leg. He had forgotten the other man and now he lay helpless on the sand hearing only the heavy panting of his own breath and the jeering of the festive crowd keeping their Roman holiday. The man with the sword approached him and looked at him carefully as if measuring where best to strike. But there was to be no single stroke or *coup de grâce* and he was hacked to death while the crowd laughed and urged his tormentors on.

This was the life and death of a hippo in the amphitheatre in Roman London.

He was puzzled by the loud noise of the rifle shots and why his mother simply crumpled down, two bloody holes having suddenly appeared in her side. He went to her but then ran away in confusion to hide in the reeds as he had been taught to do when danger threatened. He could hear them coming and although he tried to escape he felt a terrible pain as he was speared by a barbed harpoon. He was pulled into the boat and struggled so much that it capsized, but he was too weak to fight for long and he was dragged from the water to a waiting cage. As he was driven away he could see a dark and struggling crowd of vultures already tearing at his mother's steaming body.

But this is the beginning of Obaysch's story, and if I told more of it now this book would be soon over.

Introduction: Why Obaysch?

What is remarkable about Obaysch was that he was the first live hippopotamus (*Hippopotamus amphibius*) to be seen in England since Roman times. Indeed, he was the first live hippo to be seen in Europe.

There had once been plenty of hippos in Britain. During the second inter-glacial period, which ended about 125,000 years ago, they ranged across the island inhabiting an environment sustained by climactic conditions roughly equivalent to those of modern day Africa and enjoying co-existence with many other animals that we now think of as exclusively African. Perhaps less enjoyably – as the size and strength of hippos make them more or less immune to predators once they reach maturity – they co-existed with the humans who were also enjoying the warm climate and who retreated south, keeping ahead of the ice as it inexorably covered the savannahs of lowland Scotland and the tropical forests of Middlesex. Hippos make good eating and hippo bones are not uncommon finds. Stand in Trafalgar Square, for example, and you are standing where hippos once wallowed. Catch a train from Charing Cross and a few feet beneath the platform will be the remnants of a landscape that was once rich in hippos.

By Roman times things had warmed up again and pretty good red wine was being made from grapes grown in vineyards at Vinovia near Bishop Auckland in England's now decidedly chilly north-east. It is only probable that hippos were imported into Roman Britain for gladiatorial displays. Certainly they were imported into Rome and the Emperor Commodus, a keen amateur gladiator, once killed five in a single session.[1] In 1850 a note about Obaysch in the *Glasgow Herald* picked this up and made it the occasion for one of those 'how different, how very different, from the life of our own dear Queen' comments that make the Victorians so rewarding to study:

> Commodus on one occasion exhibited five; and descending into the arena butchered some of these wretched beasts with his own hand. Queen Victoria, accompanied by her consort and their children, the hopes of Britain, can now graciously look upon the unmolested creature.

It is hard to comprehend the scale of the Roman games as far as animals were concerned. Thousands of creatures of every species were imported into Rome and marched through the streets to be slaughtered to liven up a holiday or reinforce a triumph. Nine thousand were killed in the games that inaugurated the Colosseum. There were certainly animal shows in Roman London, where the amphitheatre reproduced the pleasures of Rome on a provincial scale, but we do not know for sure whether or not hippos were ever floated down the Thames to titillate the crowds. The teenaged Emperor Gordian III had a huge menagerie that included at least three hippos, although there is some doubt as to whether this was for his amusement, like a private zoo, or a holding pen for a planned *venatio*, a public entertainment in which wild animals were released, 'hunted', and killed. The slightly mysterious *Historia Augusta* tells us

---

1   J.M.C. Toynbee, *Animals in Roman Life and Roman Art* (Baltimore: Johns Hopkins University Press, 1996), pp. 128–130.

that the wealthy usurper Firmus, who sounds very much like a real life Trimalchio, had a collection too and would take the great risk of riding on his hippo and his ostrich (which must have been a strong animal, as Firmus was a famously big man). It would be nice to think that instead of merely being kept for slaughter in the arena hippos might have been kept for decoration like the tamed lion with a gilded mane mentioned by Seneca or the 300 Mauretanian ostriches that Gordian I had dyed vermillion. I wonder if Lord Berners knew about that when he dyed his flock of doves? Probably he did.

There was other knowledge of hippos. Soldiers who had served in Egypt or were adepts of the cult of Isis may well have had an interest in Taweret, the hippo god.[2] Certainly coins of the Empress Otacilia Severa stamped with the image of a hippo were circulating and have been found in at least one buried hoard in the British Isles. Did their original owners know what the image depicted, or did they just see yet another monster?

After the final surviving hippo breathed his or her last on the developing tundra or in the arena, Europe was not to see another live hippo until 1850 and the arrival of little Obaysch. There are a number of reasons for this. Firstly, hippos are very big and very fierce animals. They are, therefore, very hard indeed to catch alive. They live a long way from Europe and, in the absence of a unified state such as the Roman Empire, mounting an expedition to capture

---

[2] P. Germond, in *An Egyptian Bestiary* (London: Thames & Hudson, 2001), pp. 136 and 172, briefly surveys Egyptian hippo cults and representations of hippos. The blue ceramic faience models that were common in ancient Egypt are common again today. Both the British Museum in London and the Metropolitan Museum of Art in New York have reproductions for sale. The Met sells copies of the famous hippo statuette known as 'William' (https://bit.ly/2iuNWd1), and even offers reproductions of William as a soft toy. In 2017–18 the Met staged an exhibition called 'Conversation between Two Hippos' in which William and related artefacts were displayed alongside a ceramic hippo potted by Carl Waters in 1936. The faience hippos found in Egyptian graves almost invariably have three of their legs broken. This was done, it is believed, to hamper them when they were re-animated in the afterlife. This speaks rather eloquently to the violent nature of hippos.

one would have required a very significant expenditure of money and people. And then, why would you want one? The royal and aristocratic menageries had not assumed the importance that they did later on and there were no public or private zoos. A king might like a lion or a few bears to reinforce his sense of status. The odd elephant or giraffe might be entertaining. A rhinoceros made a very rare visit. But the hippopotamus remained beyond the reach of desire and none were dragged across the Mediterranean to entertain the princes and popes (although there is an ambiguous report of one being shipped to Amsterdam in the early seventeenth century, possibly the one that, by 1697, was being exhibited – stuffed – at the University of Leiden).[3]

In 1615 Rubens summoned up his very considerable imaginative forces and produced a significant painting depicting a hippopotamus and crocodile hunt for the Elector Maximilian I of Bavaria. This was part of a series of paintings of hunts – the others were lion, bear and boar – for the Schleissheim Palace. In the painting a hippo stands snarling in the middle of the composition with a crocodile below him, while Arabs on horse and foot and savage dogs swirl round it. The hippo's head is garnished with fearsome teeth and forms the centre of a whorl of energy in which the chaos he represents is subdued by the forces of humanity and the animals it has trained for its own purposes. The image of the

---

[3] See D. Hahn, *The Tower Menagerie* (London: Pocket Books, 2004); G. Ridley, *Clara's Grand Tour* (London: Atlantic Books, 2004); C.E. Jackson, *Menageries in Britain 1100–2000* (London: The Ray Society, 2014); C. Grigson, *Menagerie* (Oxford: Oxford University Press, 2016); C. Plumb, *The Georgian Menagerie* (London: I.B. Tauris, 2015); S.A. Bedini, *The Pope's Elephant* (Manchester: Carcanet, 1997); M. Belozerskaya, *The Medici Giraffe* (New York: Little, Brown & Co., 2006); M. Allin, *Zarafa* (London: Headline, 1998); J. Simons, *The Tiger That Swallowed the Boy* (Faringdon, UK: Libri, 2012); and G. Mabille and J. Pieragnoli, *La Ménagerie de Versailles* (Arles: Éditions Honoré Clair, 2010). L.C. Rookmaaker, in *The Zoological Exploration of Southern Africa 1650–1790* (Boca Raton, FL: CRC Press, 1989), p. 291, mentions the Leiden hippo and other early stuffed hippos exhibited in the Hague.

Introduction: Why Obaysch?

hippo is anatomically very accurate and almost certainly derives from Rubens' viewing of two mounted hippo carcasses that had been brought to Italy in 1601 by the surgeon Federico Zerenghi, who dissected them and stuffed them.[4] Whatever the physiological shortcomings of Zerenghi's taxidermy, Rubens was able to imagine the animal in real life – much as George Stubbs was able to do when he produced the first European painting of a kangaroo some 200 years later. His painting was the first time in many years that a hippo had found its way into European art, and also the last time for nearly 200 years that a reasonably accurate image of the animal would appear.

In addition, the Egyptians themselves, who had in this pre-colonial period a tight monopoly on the hippos – who lived much further down the Nile than they do now and were much more plentiful – appear to have been quite reluctant to allow them to be interfered with. They had worked out a practical mode of living by which the natural antagonists, riparian Nile farmers and hippos, could co-exist and appear to have been quite protective of their unwieldy neighbours. Certainly there is at least one recorded episode from the early eighteenth century where locals in the vicinity of Rosetta drove off a French explorer who was attempting to capture a hippo.[5]

Changing social and economic conditions and a much revised geopolitical map together with a new spirit of scientific discovery made it almost inevitable that by 1848 or 1849, when Obaysch was born and taught to graze happily with his mother on the banks of the Nile, the eyes of Europe would turn again to hippos and desire to see them alive. And now there were the new scientific

---

4   G.L. Leclerc (Comte de Buffon) tells us about Zerenghi's hippo in his 1785 *Natural History of Quadrupeds*. I have referred to a nineteenth-century English translation (Edinburgh: Thomas Nelson & Peter Brown, 1830), vol. 2, pp. 326–327.
5   A. Mikhail, *The Animal in Ottoman Egypt* (New York: Oxford University Press, 2014), p. 169.

zoological gardens to put them in. After the foundation of the first modern establishment, the Jardin des Plantes in Paris in 1792 (this was the date of the agreement to set up the institution but, arguably, the Jardin itself didn't properly exist until 1794), scientific and educational zoos were becoming an expected feature of the new cities of industrialised and industrialising Europe, the United States of America and Australia.

The history of Obaysch has been sketched several times before. Firstly in Nina Root's seminal article on Victorian 'Hippomania' and, more recently, in an article by Andrew Flack, in Narisara Murray's interesting doctoral dissertation, and in Tom Quick's excellent and as yet unpublished Master's thesis. Root, Flack, Murray and Quick each explore the tale of Obaysch from different and multiple perspectives. They show how an industry of cultural production grew up around him. They look at the various discourses (scientific, aesthetic and colonial) that figured him. They set Obaysch squarely into his time as a signifier of a set of peculiarly Victorian contexts: sometimes intersecting and sometimes floating free.[6]

In addition, almost everyone who has looked at the history of London Zoo, from Wilfred Blunt to Takashi Ito, has commented on Obaysch and, especially, on his celebrity and popularity not only as a new animal but as the first representative of a new class of animal: the ' star'. This was a new way of configuring a creature and it would not only define the zoo's new business model and growth strategy for many years to come but also guarantee its continuation as a place of both education and recreation in London until well

---

6   A.J.P. Flack, '"The Illustrious Stranger": Hippomania and the Nature of the Exotic', *Anthrozoös* 26 (2103), pp. 43–59; N. Murray, 'Lives of the Zoo: Charismatic Animals in the Social World of the Zoological Gardens, 1850–1897' (unpublished Ph.D thesis, Indiana University, 2004); N. Root, 'Victorian England's Hippomania', *Natural History* 103 (1993) pp, 34–39; T. Quick, 'Interpretations of London's Zoological Hippopotami' (unpublished M.Sc. thesis, London Centre for the History of Science, 2007).

into the twentieth century.[7] Finally, commentators on the history of zoos and the place of exotic animals in Victorian culture more generally, such as Helen Cowie, Caroline Grigson, Christopher Plumb, Christine Jackson, Peta Tait and Sarah Amato (and several of my own works have explored this too) have generally paused for a longer or shorter moment on the unique place occupied by Obaysch as the 'first hippopotamus seen in Britain (or Europe) since Roman times'.[8]

More generally scholars like Randy Malamud and Nigel Rothsfels have placed the zoo within the context of a reading of human–animal relationships defined, especially, by empire and by species domination.[9] The present work shares many of these ways of seeing and the similarities and differences between my approach and theirs will be further explored below. It is also the case that neither Malamud nor Rothsfels loses sight of the specifics of the animal experience in captivity and the harms that are done to animals through that experience. This is absolutely central to

---

7 J. Barrington-Johnson, *The Zoo* (London: Robert Hale, 2005); W. Blunt, *The Ark in the Park* (London: Hamish Hamilton, 1976); A. Desmond, 'The Making of Institutional Zoology in London 1822–1836', *History of Science* 23 (1985), pp. 133–185; T. Ito, *London Zoo and the Victorians* (Woodbridge: Boydell for the Royal Historical Society, 2014).

8 S. Amato, *Beastly Possessions: Animals in Victorian Consumer Culture* (Toronto: University of Toronto Press, 2015); H. Cowie, *Exhibiting Animals in Nineteenth-Century Britain: Empathy, Education, Entertainment* (London: Palgrave Macmillan, 2014); P. Tait, *Fighting Nature: Travelling Menageries, Animal Acts and War Shows* (Sydney: Sydney University Press, 2016); C. Jackson, *Menageries in Britain 1100–2000* (London: The Ray Society, 2015); C. Grigson, *Menagerie: the History of Exotic Animals in England* (Oxford: Oxford University Press, 2015); C. Plumb, *The Georgian Menagerie* (London: I.B. Tauris, 2015); J. Simons, *The Tiger That Swallowed the Boy*.

9 R. Malamud, *Reading Zoos: Representations of Animals and Captivity* (New York: New York University Press, 1998) and N. Rothfels, *Savages and Beasts: The Birth of the Modern Zoo* (Baltimore: Johns Hopkins University Press, 2008) are both foundational to the study of zoos in the context of imperial expansion and domination.

my own sense of purpose in pursuing animal studies at all and especially in looking at individual animals in captivity.[10]

So what is to be said about Obaysch that hasn't already been said? And why write a full scale – albeit relatively compact – book about a single animal?

There are a number of motivations.

The first is that the excellent scholarship I have already mentioned is often, essentially, scholarship about Obaysch as a cultural token. What I would like to do, at least in part, is to treat Obaysch as an actor in his own life and to write his biography. I want to treat him as a living being and not simply a sign. This is a strategy I have pursued before in my book on Top, the wombat owned by the Pre-Raphaelite artist and poet Dante Gabriel Rossetti.[11] Top, however, lived only a few months and the details of his life were correspondingly few and required much reconstruction. In contrast, Obaysch lived for at least twenty-eight years. He was the centre of attention in the Zoological Gardens. There are many rich and varied sources from which we can piece together the textures of his daily life: letters, newspaper and magazine articles, minutes, proceedings, reports and scientific papers of the fellows and managers of the Zoological Society, travel books, memoirs, cartoons, drawings, paintings, works of fiction and magnificent photographs. Obaysch is as well documented as any Victorian celebrity – perhaps better documented than some – and all the materials for a conventional biography are in place and not particularly hard to find. Obaysch had what might, anthropomorphically, be described as a family or, in hippopotamine terms, a bloat. He had human companions and visitors, some of whom, such as Queen Victoria and Prince Albert,

---

10 For example, see L. Marino, S. Lilienfeld, R. Malamud, N. Nobis and R. Broglio, 'Do Zoos and Aquariums Promote Attitude Change in Visitors?', *Society and Animals* 18 (2010), pp. 126–138; and N. Rothfels, 'How the Caged Bird Sings: Animals and Entertainment', in K. Kete (ed.), *A Cultural History of Animals in the Age of Empire* (Oxford: Berg, 2011), pp. 95–112.
11 J. Simons, *Rossetti's Wombat* (London: Middlesex University Press, 2008).

Introduction: Why Obaysch?

Lord Macaulay, Charlotte Brontë and Charles Dickens, were (or in Brontë's case, was becoming) extremely famous.

So there is a rich life and interesting life to be written here, and if we are to take the role of scholarship in the pursuit of animal welfare seriously we should try, where we can, to put the animals at the centre of their own narratives. We should try to give them at least a symbolic form of agency. We cannot reconstruct their feelings or perceptions, of course, and nor should we think we can – except perhaps in the attempt to empathise with their sufferings. But we can endeavour to see them as more than an effect of signification, as more than a term in a cultural value chain, or simply as a material intervention in a moment of social, economic or scientific change. We can, for example, acknowledge that Obaysch had a life of perhaps at least a year before he was captured, before he was named and before he became a celebrity. When we consider this and contrast the life of a wild hippopotamus with that of a caged one, however loved, cosseted and pampered, perhaps we can begin to understand just how cruelly Obaysch was treated and how shocking the incarceration of free-ranging animals can be.

In fact, the story of Obaysch is a story of misery and pain. I have explored more sources than anyone previously and the more I read the more I realised that Obaysch's story was layered in unexpected ways. As I triangulated different accounts, a narrative different from the received one began to reveal itself. This suggested that Obaysch, far from being the pet of the zoo, was a long-term problem to which there were few answers. What was also revealed was an information and media management strategy that concealed the extent of the issues from Obaysch's extremely loyal public.

The second motivation is that it is all too easy to understand Victorian zoology and especially the growth of the Zoological Gardens and the acquisition of wild animals from Africa in particular solely through the lens of a very specific reading of imperial history. By this argument – which is found in complete or fragmentary form in almost all writings on this topic including my own – the capture, killing, study, display, stuffing, mounting, preserving, turning into furniture, jewellery or clothing of wild animals were the epiphenomena of the logic of domination that made empire possible. The colonisation of people is, in this line of thought, seen as being reinforced by a colonisation of their environment.[12] For example, big game hunting can readily be shown to be a master term in a chain of signification which sets out the superiority of the European not only over the non-European people but also over the non-European animals.[13] It is, to use another set of well-known political terms, the hegemonic counterpart of domination or even a form of display which sits interestingly between the hard-power projections of military conquest and the soft-power projections of trade and education. One can see this logic at work, for example, in the ambiguity by which ethnographic displays ranging from the stirring entertainments of Buffalo Bill's

---

12 See J. MacKenzie (ed.), *Imperialism and the Natural World* (Manchester: Manchester University Press, 1990); H. Ritvo, *The Animal Estate* (Cambridge, MA: Harvard University Press, 1987); L. Schiebinger, *Plants and Empire* (Cambridge, MA: Harvard University Press, 2004). Animals can themselves be agents of colonisation and attention should be paid to the impact of non-indigenous animals on ecosystems as part of the colonising process. See V.D. Anderson, *Creatures of Empire* (New York: Oxford University Press, 2004); A. Crosby, *Ecological Imperialism* (Cambridge: Cambridge University Press, 2004); C. Forbes, *Australia on Horseback* (Sydney: Macmillan, 2014); and S. Nicholls, *Paradise Found: Nature in America at the Time of Discovery* (Chicago: University of Chicago Press, 2011).

13 P. Boomgard, *Frontiers of Fear* (New Haven, CT: Yale University Press, 2001); J. MacKenzie, *The Empire of Nature* (Manchester: Manchester University Press: 1988); P. Verney, *Homo Tyrannicus* (London: Mills & Boon, 1979). Of course, the logic of domination played out in hunting is not only about imperialism. It is also about the dominance of local rulers and speciesism. The Indian *shikar* is a good example of this and went on before, alongside and after the British tiger hunters had left. See J.C. Daniel and B. Singh, *Natural History and the Indian Army* (Mumbai: Oxford University Press for Bombay Natural History Society, 2009). The Mysore taxidermists Van Ingen and Van Ingen stuffed or preserved some 43,000 leopards and tigers, some for visiting hunters but mostly for native princes. See P.A. Morris, *Van Ingen and Van Ingen* (Ascot, UK: MPM Publishing, 2006).

Introduction: Why Obaysch?

Wild West Show (which was at one level quite literally a travelling prisoner of war camp containing Sioux captured in the aftermath of the Ghost Dance uprising) or the imperialistic epic spectacle shows such as Fillis's Savage South Africa through to the sad dancing of Tambo and his Indigenous Australian colleagues as they toured the United States. We see it in the native villages set up alongside the animal enclosures in various zoos around the world and which were incorporated in the display of conquered nature. An alarmingly thin and porous line can sometimes be drawn between the native animals and the native people. Melbourne Zoo, for example, had a native village where people could watch Indigenous Australians.[14] This thin line can most notably be seen, perhaps, in Carl Hagenbeck's Hamburg *Tierpark* which, by the twentieth century, had become, it feels, almost more of an ethnographic exhibition than a zoological facility.[15]

There is, of course, much truth in this postcolonial argument, especially as it is articulated in Harriet Ritvo's seminal work *The Animal Estate* (behind which stands Sir Keith Thomas's *Man and the Natural World*) and much of her subsequent writing and in Alfred Crosby's much less frequently cited *Ecological Imperialism*, and its operation can be demonstrated time and time again.[16] However, as more and more people have taken up the idea of studying history and culture by means of looking at animals and human–animal

---

14   See P. Blanchard et al., *Human Zoos: Science and Spectacle in the Age of Empire* (Liverpool: Liverpool University Press, 2008); B. Bridger, *Buffalo Bill and Sitting Bull: Inventing the Wild West* (Austin: University of Texas Press, 2002); N. Montagnana-Wallace, *150 Years Melbourne Zoo* (Thornbury, VIC: Bounce Books, 2012); R. Poignant, *Professional Savages: Captive Lives and Western Spectacle* (New Haven, CT: Yale University Press, 2004); J. Simons, *The Tiger That Swallowed the Boy*, pp. 81–98; P. Tait, *Fighting Nature*.
15   E. Ames, *Carl Hagenbeck's Empire of Entertainments* (Seattle: University of Washington Press, 2009).
16   A. Crosby, *Ecological Imperialism*; Ritvo, *The Animal Estate*; K. Thomas, *Man and the Natural World: Changing Attitudes in England 1500–1800* (Oxford: Oxford University Press, 1983).

relations it increasingly seems to me that the subtleties of the postcolonial argument as expounded by Ritvo have been lost and it has too often become an almost Manichaean scheme imposed on a much more complex set of practices and concepts. The British Empire and the currents of thought and action that flowed through and round it were really not that simple. Who would have thought, for example, that the public outrage which blew up around the fact that English (white) women were allowed in as paying customers to Fillis's Zulu *kraal* where they could mingle with nearly naked Zulu (black) men was motivated not by the fear that the savages might molest the women or, at the very least, put unladylike thoughts into their delicate heads but rather that the women might corrupt the savages? And yet that is the case, and the argument for restricting access to the *kraal* on these grounds even raged in the Parliament of the time.[17]

It is also far from the case that the desire to capture and display wild animals was a purely European one that can be linked simply to a causal chain that leads back to imperialism (although such links do exist). For example, during the period (say 1851–1900) of expansion of zoos in the United States and Europe and colonies such as Australia, Canada and South Africa (which may be counted as extensions of Britain for this purpose), zoos were also developing elsewhere. In India (including that region that later became Pakistan) the traditional princely menageries were now supplemented by seventeen new city zoos. Zoos were established in Japan, Indonesia, Viet Nam, Hong Kong and Singapore. In South and Central America Argentina, Brazil, El Salvador and Guatemala all built zoos. Of course, some of these may be explained by the same logic that is used to explain European zoos and seen as a simple extension of colonial rule. However, that is not so possible for zoos in India, even though it was a colony. Nor for Japan, which was fiercely independent although fast learning from the West about industry and commerce. In India

---

17 Simons, *The Tiger That Swallowed the Boy*, pp. 84–85.

Introduction: Why Obaysch?

the zoos were not there solely to serve the very small European communities: they had scientific and educational missions for Indians. There has not yet been a proper study of zoos in India under the Raj but when there is I think we will find that underlying all the process of colonisation, decolonisation and postcolonisation is a more fundamental process of speciesism. All zoos, whether in the imperial heartlands, in colonies or in countries which are not readily aligned with either of those categories, respond to an analysis rooted in the categories of anthropocentric species domination.[18]

Insufficient attention is often paid to the contexts in which imperialism operated. This lack of attention chiefly manifests itself in two forms. The main one is a failure to note that the most significant imperial project of the post medieval period was not the British Empire nor the French, Belgian, Russian or German empires. It was not the growing global power of the United States. It was not Christian and it was only partially European. I am referring to of course to the Ottoman Empire (one of three so-called Islamic Gunpowder Empires, the other two being the Mughal and the Persian) which was much longer lived than any of the Western European empires and more extensive than most of them. In Egypt, for example, the Ottomans ruled in some form or other for 397 years, which may be compared to the sixty-eight years that the British spent in Uganda, or indeed to the 200 years of the British Raj (in contrast, the Mughal colonisation of India lasted over 300 years).

British activities in northern and central Africa were almost always carried out against a backdrop of the Ottoman Empire, either

---

18 There are some preliminary studies of this fascinating topic and the related topic of botanical gardens: S. Walker, A. Pal, B. Rathanasabapathy and R. Manikam, 'Indian Zoological and Botanical Gardens: Historical Perspective and a Way Forward', *BGjournal* 1 (2004), https://bit.ly/2DoiwiH; V. Kisling, 'Colonial Menageries and the Exchange of Exotic Faunas', *Archives of Natural History* 25 (2000), pp. 303–320; and C.V. Hill, 'Colonial Gardens and the Validation of Empire in Imperial India', *Journal of South Asian Studies* 1 (2013), pp. 139–145. See also K. Rookmaker, 'The Royal Menagerie of the King of Oudh', *Back When, and Then* 2:1 (August 1997), p. 10.

in Ottoman territory such as Egypt or in countries such as the Sudan and what was to become Uganda. Here the colonial pressures of the Ottoman Khedivate of Egypt, combined with the activities of Arab and African slave traders, had comprehensively smashed local cultures. It is not possible to understand some of the motivations of British colonisation in sub-Saharan Africa without understanding the very specific conditions created by the Ottoman sphere of influence and the Ottoman trade routes.[19]

Another issue is lack of attention to non-European interaction with animals. There are rare exceptions to this tendency, such as Peter Boomgaard's work on tigers in Indonesian, Malaysian and Singaporean culture, Alan Mikhail on animals in the Ottoman Empire, Brett Walker on wolves in Japan, David Rockwell on the native American relationship with bears, Robert Kenny on Indigenous Australians and sheep, and Greg Bankoff and Sandra Swart's collection of essays on horses in South-East Asia and southern Africa.[20] But, broadly, the scholarship within the Anglosphere concentrates on the Western empires and this can have a distorting effect on the ways in which the practices and processes of imperialism are understood.

As I shall suggest in the next chapter, the specific facts of Obaysch's capture can, in many ways, be attributed to the complex interplay of culture and political motivation between the British and Ottoman Empires in Africa. It might even be read as an early sign of the decline of Ottoman power in the face of the growing influence of Britain.

---

19 Tim Jeal's works, notably *Livingstone* (London: Pimlico, 1993), *Stanley* (London: Faber & Faber, 2011), and *Explorers of the Nile* (London: Faber & Faber, 2011), offer an excellent and readable account of these matters.
20 Boomgard, *Frontiers of Fear*; Mikhail, *The Animal in Ottoman Egypt*; B. Walker, *The Lost Wolves of Japan* (Seattle: University of Washington Press, 2008); D. Rockwell, *Giving Voice to Bear* (New York: Roberts Rinehart, 2003); R. Kenny, *The Lamb Enters the Dreaming* (Melbourne: Scribe, 2007); G. Bankoff and S. Swart, *Breeds of Empire* (Honolulu: University of Hawai'i Press, 2007).

Introduction: Why Obaysch?

What I am trying to argue here is not that the broad sweep of the postcolonial reading of the zoo and related spectacles is wrong, but that the imposition of a too narrowly framed argument about empire and its aftermath is not sufficient fully to explain what was going on when a hippopotamus was shipped from Africa to London. Nor is it capable of making the narrative space necessary to understand the textures of that hippo's life as an individual animal with some agency and social relationships of his own. So although one cannot tell the story of Obaysch without acknowledging his role as a very small part in a very big imperial picture, neither can one tell it properly if one assumes too easily a facile version of the narrative of domination which I have sketched out above, admittedly very briefly and in fragmentary form, as part of another story entirely.

In many ways the story of Obaysch is precisely that of that most endearing character of postcolonial fiction: Babar the Elephant.[21] Like Obaysch, Babar sees his mother gunned down by a hunter and finds his way to a European city. There he is befriended by an old lady dressed in black, as Obaysch was befriended by Queen Victoria. He marries and has children and becomes the king of his own domain. This is also true of Obaysch, except Obaysch was not repatriated and uneasily ruled over only a small part of London Zoo. And just as Babar has counsellors and companions so Obaysch had his Egyptian keeper Hamet Cannani, Sir Charles Murray, David Mitchell, Abraham Bartlett and various other members of the

---

21 The Babar books were written by the French author Jean de Brunhoff between 1931 and 1941 and the series was continued by his son Laurent between 1948 and 2014. The image of Babar standing crying beside the dead body of his mother reminds me painfully of the photograph taken of the baby elephant who was to become Jumbo (not the famous Victorian animal but another elephant: the first East African elephant to be displayed in Germany) standing hopelessly and helplessly beside the body of his dead mother after she had just been killed by the brave and noble hunter Hans Schomburgk, an image reproduced in his book *Wild and Wilde im Herzen Afrikas* (Berlin: Deutsche Buchgemeinschaft, 1925).

Zoological Society staff, and the irrepressible naturalist and sometime surgeon to the Life Guards, Frank Buckland. Babar's story is now seen as a malevolent outcrop of postcolonialism and has been removed from the shelves of some public libraries in Britain. Interestingly, it is not seen – as it might easily be – as a narrative which tells of the cruel exploitation and capture of exotic animals as an effect of empire and a fantasy of how this might be recuperated. In trying to tell Obaysch's life story I am attempting to nuance the argument and to find more in the twenty-eight years life of a caged hippopotamus who was captured – kidnapped one might say – from the wild than a simple metaphor for the evils of imperialism. I will not accept that Obaysch, a real animal who lived a real life in the Zoological Gardens, is simply to be seen as a fragment of collateral damage washed up in an obscure corner of imperial historiography.

The special status of Obaysch depends very heavily on his being the first hippopotamus in Britain (or Europe) since Roman times. There are other claimants to this title, but I believe that these claims are all based on misreading or misunderstanding of the sources. Although you will find histories of eighteenth-century and nineteenth-century menageries which claim that live hippos were exhibited in London before the arrival of Obaysch in 1850, I do not believe that this was the case; I believe that Obaysch really was the first hippo to be seen in Britain in nearly 2000 years – or 150,000 years if you don't think the Romans imported any.

If we look at European specimens first, we will find that there are two claimants. The first is the hippo allegedly shipped into Amsterdam. I do not believe that this was a live hippo. Either it was already dead and had been shipped to be preserved, or it was a different species of animal entirely.[22] The reason for my

---

22 In fact I think this hippo is the skin which was imported into Holland in 1670 and given to the University of Leiden. William Dampier mentions it in his *Voyages and Descriptions*, Volume 2 (James and John Knapton: London, 1699), pp. 103–104, and it was described at length in Jean Nicolas Sébastien Allamand's note to the description of the hippopotamus in Buffon's *Natural History*

Introduction: Why Obaysch?

skepticism is simply the absence of documentation (which is very thin) and of imagery (which appears to be non-existent). It seems to me almost impossible that there would have been a live hippo in seventeenth-century Holland, where every second person seemed to be wielding a paintbrush and every first person was commissioning them to do it, without more being known about it. It has, quite simply, left no credible trace. A stuffed hippo might have been a curiosity, but no more curious that the wealth of strangeness which was pouring into Western Europe at that time. But a live hippo would have been just as much of a sensation as Obaysch was 250 years later. In addition, by the mid nineteenth century there was some very good scholarship on the history of zoos and menageries and although it was, of course, very much in the interests of the Zoological Society of London to make an exceptional claim for Obaysch's rarity (it was his unique selling point, as we might say today), and although everyone else wanted to believe that Britain had won the race to acquire a live hippo (even the astonishingly acquisitive P.T. Barnum didn't manage to get a hippopotamus into the United States until 1861) in practice, evidence that the Amsterdam hippo predated Obaysch would surely have emerged if it existed. Not least because some of the

---

(London: W. Strachan and T. Cadell, 1785), volume 6, pp. 302–304). By this time it had been mounted on iron rings to give some verisimilitude. Zacharias von Uffenbach visited it on 16 January 1711 and noted that '*man muss sich verwundern das man es hierher gebracht, wegen seiner gewaltige Grosse*' ('people must wonder to themselves that someone brought it here because of its enormous size'). *Merkwürdige Reisen durch Niedersachsen, Holland und Engeland* (Noteworthy travels through Lower Saxony, Holland and England) (Frankfurt and Leipzig, 1754), volume 3, pp. 405–406. This was not the only hippo skin to be seen in Europe in the eighteenth century. There was one in the Prince of Orange's cabinet of curiosities described by Jacob Klöckner in another supplement to Buffon (pp. 305–314). Another was in the Prince de Condé's cabinet at Chantilly, where he also kept a significant menagerie. This was seen by Samuel Johnson, who thought it was a fake on account of its tiny size, on 2 November 1775. J. Boswell, *The Life of Samuel Johnson*, edited by R. Ingpen (London: George Bayntun, 1925), volume 1, p. 551.

best scholarship in this field was French: there had been, since the French Revolution, low-level but persistent competition between France and Britain in the matters of zoo-keeping and animal collection.[23]

A more difficult case is the stuffed hippo now in La Specola museum in Florence. The local tradition is that this animal was originally part of Cosimo III's collection in the Medici menagerie in the Giardino del Boboli, where it is supposed to have lived in one of the fountains in the very late seventeenth century. The sole reason for accepting this story is the local belief in it. This should not be lightly discounted, in spite of an absence of further compelling evidence. Reasons to doubt the story, however, are twofold. Firstly, we know that even in the early period of the Boboli menagerie the display was divided between live animals and realistically stuffed ones. In such early menageries, well into the eighteenth century, painting and statues often supplemented or even stood in place of live animals. Over the years what had always been a stuffed animal may have been re-imagined as a live one. In this case, there are clear chain marks on the skin of the stuffed hippo, but these could have been made while the animal was captive in Africa. Secondly, in spite of their size, enormous power and occasional terrifying ferocity, hippos are actually quite delicate creatures and, as we shall see in the case of Obaysch, they do not do well in captivity unless they have ample swimming space and good food of the right kind and in copious quantities. It seems to me highly unlikely that a hippo could live for very long in a fountain. It has to be admitted that a hippo really might have lived a brief life as a plaything of the later Medicis – but it is highly unlikely and so, in spite of my respect for the local

---

23   G. Loisel, *Histoire des Ménageries de l'Antiquité à nos Jours*, 3 volumes (Paris: O. Doin et Fils, 1912) remains, in my opinion, a work of scholarship yet to be surpassed and makes no mention of this hippo. Nor does V. Kisling (ed.), *Zoo and Aquarium History* (Boca Raton, FL: CRC Press, 2001), which is the best and most comprehensive history of zoos to appear in recent years.

story, I would have to conclude that this hippo was always a model of taxidermy and never a live specimen.[24]

There was also, it appears, a hippopotamus sent as a gift from the Pasha of Egypt to the Emperor of Austria, who kept a significant menagerie at the Schönbrunn Palace, in 1818. There is no evidence that it ever arrived in Europe; presumably it died en route, as did so many animals during the march out of Africa. The architect Charles Barry heard of it while he was travelling in Egypt and noted it down in his journals, but that is all we know of it. It has to be assumed that it did not leave Africa or that, if it did, it died somewhere on the Mediterranean. Perhaps, like an unfortunate hippo documented in 1814, it became so difficult to manage on the sea voyage that it was thrown overboard like the rotten meat in Conrad's *Heart of Darkness*.[25]

Moving from continental Europe to England, the first claim that is made for a live hippo is for the one allegedly displayed in the wonderfully eccentric Sir Ashton Lever's wonderfully eccentric showcase, the Holophusikon. This was a museum of natural history and ethnography opened in a mansion in what is now Leicester Square in central London in 1775. In effect it was a cabinet of curiosities the size of a large building, and could be inspected by members of the public for the handsome sum of half a guinea. This equates to at least £6000 today, so it was not exactly a cheap day out. As we shall see, admission prices have an important part to play in Obaysch's story too. The animal in the Holophusikon is sometimes thought to have been a live hippopotamus. But even if we were seriously to entertain the thought that one could keep such a big animal reasonably healthy in a London townhouse, the fact is that it was stuffed and took up most of the space in a room on the first

---

24 See L.E Thorsen, 'The Hippopotamus in Florentine Zoological Museum "La Specola"', *Museologia Scientifica* 21 (2004), pp. 269–281. M. Morton (ed.), in *Oudry's Painted Menagerie* (Los Angeles: Getty Publications, 2007), tells the story of a wholly artificial collection of animals from the eighteenth century.

25 A. Barry, *The Life and Works of Sir Charles Barry* (London: James Murray, 1867).

floor.[26] A representation of the head of this specimen can be seen in a watercolour by William Samuel Howitt. Howitt had never seen a hippo himself. In fact many of his wildlife paintings were from stuffed specimens. But he managed a representation by combining a study of the head from the Holophusikon's hippo and a copy of the painting of the hippopotamus by Samuel Daniell, who had travelled extensively in Africa and who drew his animals from life. This watercolour, published as a plate in Daniell's *African Scenery and Animals* (1804), offered one of the more realistic images of a hippo then available to the European reader. Interestingly, although Howitt's image does quite well for the body (where he relied on Daniell), he departs from reality lamentably with regard to the head (which he took from the stuffed specimen), showing two boar-like tusks projecting upwards on each side of the mouth and two tusks like an elephant's projecting forward. This leads one to wonder whether the example in the Holophusikon had in some way been 'improved' to make it look more like people's general idea of a hippo. The question of the hippo's tusks will be discussed in more detail below.

Confusion between stuffed and live animals in early menageries is not uncommon. For example, some visitors to the Great Exhibition complained that the advertised menagerie they had been looking forward to was in fact entirely composed of stuffed animals. It was scant consolation to know that the hand of the great John Gould himself had been at work on many of them, such as the duck-billed platypus. The comparative rarity of hippos, stuffed or live, can be gauged by the fact that when a baby hippo was exhibited (stuffed) at the Egyptian Hall in Piccadilly in 1827, it merited special mention in the *Times*. The fact that this specimen was sufficiently rare to merit such notice seems to me further evidence that, in the

---

26   A.L. Kaeppler, *Holophusicon* (Altenstadt, Germany: ZFK Publishers, 2012) offers a thorough study of this engaging institution. Grigson, *Menagerie* and Jackson, *Menageries in Britain 1100–2000* also discuss the Holophusikon.

Introduction: Why Obaysch?

early nineteenth century, there had been no live hippos recently in London.

A slightly more difficult case is the hippopotamus allegedly to be seen in the Exeter 'Change menagerie. This was one of the most important of the early menageries. When the Zoological Gardens were founded in 1823 there was significant dialogue between the naturalists of the Exeter 'Change and the Zoological Society. A durable and oft-cited piece of evidence for this hippopotamus comes from an amusing passage in Lord Byron's journal for 14 November 1813, in which he speaks of visiting the Exeter 'Change 'two nights ago' and seeing a 'hippopotamus, much like Lord Liverpool in the face.' Admission to the menagerie normally cost one shilling, but to go in at night to see 'the tigers sup' cost half a crown (two shillings and sixpence – equivalent to £460 today). It is worth noting that although there is a view that menageries like the Exeter 'Change attracted a crowd of working-class drunks to wonder at and mock the animals, the evidence of admission prices shows that this could not have been the case. Although such menageries appear on occasion to have been crude and rowdy places by our standards, this is simply an effect of things being done differently in the past. The upper class has always provided its own supply of badly behaved drunks who were perhaps, being less easily cowed, harder to control than the odd dustman who had been on the beer and was determined to fight a crocodile. One can easily imagine Byron and his cronies, flushed with claret, hock and port, cheering on the tigers as they tore at bleeding hunks of beef.

But: if Byron really did see a hippo, then the animal predated Obaysch by some thirty-eight years. This so worried one J.S. Warden that in 1855 he submitted a 'minor query' to *Notes and Queries*, pointing out that:

> there can be no doubt whatsoever that the stout gentleman in Regent's Park [i.e. Obaysch] is the first of his kind that has appeared in Europe since the days of the Romans.

Warden concludes therefore that Byron must have been mistaken and speculates that what he saw was actually a tapir:

> an animal of somewhat similar habits, and the outline of whose countenance is not so utterly different from that to which it is compared [i.e. Lord Liverpool's face].

This was a pretty good guess, and had Mr Warden seen an advertisement for the menagerie placed in *Ackermann's Repository* for January 1814 he would have been able to clear up the mystery:

> The other apartments contain the Tapir, or Hippopotamus of the New World.

So what Byron saw was a tapir, but in describing it as a hippo he was using the menagerie's own terminology. It was however a New World animal (the tapir, from Malaysia, did not become commonly known in the West until the 1820s) and was a creature very different from the African hippopotamus.[27]

All in all, consideration of the available evidence leads to a clear conclusion that there were no live hippos in England before the arrival of Obaysch, and that the slightly weaker claim that he was the first live hippo to come to Europe since Roman times can certainly be strongly defended.

Obaysch's importance, however, was not only owing to his rarity. To understand his significance we also need to understand the situation of the Zoological Gardens both before and after he arrived, especially in the context of its policies about the acquisition and disply of new animals. I will suggest that the profound shift this

---

[27] J.S. Warden, 'Lord Byron and the Hippopotamus', *Notes and Queries* 12 (1855).

Introduction: Why Obaysch?

wrought in contemporary ideas about zoos has in fact created a pattern that has lasted until the present day. Most of the material in this section will be well known to specialists in zoo history. But it will be helpful to assist readers who are not familiar with the development of the London Zoo to contextualise the story of the first three hippos to live in London – Obaysch, Adhela, and Guy Fawkes – and, to a lesser extent, the other hippos which gradually found their way to Europe and beyond.

In Victorian Britain, zoos were being built at a fast rate. Leeds, Liverpool, Hull, Preston, York, Northampton, Manchester, Bristol, Rosherville, Eastham, Cardiff, Edinburgh and Glasgow (among other towns) all set up zoos between 1840 and 1900, and there were several other zoos in London.[28] Britain was the first country to have the majority of its people living in towns and cities and although the urban landscape they inhabited included a good deal of appallingly substandard housing, it also displayed a new sense of municipal pride – perhaps especially in the new industrial conurbations of the north and midlands – in magnificent public buildings, public parks and institutions designed to provide respectable and educational pastimes for the new working class. A contest for control of the economic value of leisure as well as for its social meaning was taking place and the public house and rough music hall would be, where possible, replaced by the library, the museum, the gallery, the Mechanics' Institute and, in many towns, the zoological gardens. The creation of a working class which was politically tamed and individually refined by the cultural affordances of private philanthropy and municipal benevolence, as well as by free public education, was a project which occupied much of the middle of the Victorian era and found an apotheosis of sorts in Prince Albert's own pet project, the Great Exhibition of 1851.[29]

---

28 See J. Simons, *The Tiger That Swallowed the Boy*, pp. 99–137, and V.N. Kisling, *Zoo and Aquarium History*, pp. 49–73.
29 See J.A. Auerbach, *The Great Exhibition* (New Haven, CT: Yale University Press, 1999) for a full account of this remarkable event.

However, people being what they are, the social engineering project didn't go as smoothly as might have been hoped and the extremes of the debate are articulated in Dickens's *Hard Times*, in which the social Darwinism and utilitarianism of Mr Gradgrind, akin to a secular Calvinism, are confronted by the anarchic joy of the circus owner Mr Sleery. This specific conflict of values affected the viability of many of the new institutions. Zoos, in particular, were vulnerable to closure as the envisaged crowds failed to materialise after the first excitement. When we consider the zoological garden in the Victorian era we see three kinds of establishment. The first is Regent's Park, which opened in 1828: a well-founded entity with an explicit scientific mission supported entirely by the private subscriptions of its fellows, other donations and a very high entry fee, available at first only to those who were accompanied by a fellow. It is in a class of its own as far as Victorian zoos are concerned. The second type consists of municipal zoos like that at Hull, where a collection of animals donated or purchased was available to be viewed at a relatively cheap price in the hope of educating a broad public. Sometimes this could be a very small menagerie, as at Churchtown just outside Southport or in Southport itself, where the Winter Gardens housed a collection of monkeys. Sometimes it was specialised, like the display of polar bears and sea lions at Groudle Glen on the Isle of Man. The third type consists of privately owned zoos that combined a collection of animals with other kinds of entertainment ranging from concerts, firework displays and spectacular performances to dances and archery competitions. The most spectacular example of this type was the Belle Vue Zoo in Manchester, which might well be considered the prototype of the modern theme park.

Although the municipal zoos started with high hopes and good intentions, most of them failed to thrive and they fared badly against the Belle Vue type. There were a number of reasons for this but the primary one is clear: if all that you had was a smallish collection of animals, it was difficult to persuade people to come and see

Introduction: Why Obaysch?

them more than once. In contrast, a day out at Belle Vue would probably include a tour of the zoo but also many other things as well, including a good meal in one of several outlets, each designed to appeal to a particular class and pocket. The knowledge that every visit would be different encouraged people to come time and time again. In other words, a purely educational mission could not deliver the financial sustainability that comes through repeat business. In addition, the municipal zoos had to compete with the travelling menageries, some of which were massive and highly sophisticated. These also offered a range of entertainments – such as fairground rides and side shows – that went beyond simply walking from cage to cage and peering at the animals inside. Many of the private menageries also offered limited public access.[30]

There was a great deal of money to be spent and made in exotic animals in the Victorian era but there was also a great deal to be lost. A mixture of under-capitalisation and a lacklustre product was a fatal combination for municipal zoos.

In the patrician atmosphere of Regent's Park this was not a debate that interested the fellows, but they were eventually forced to think hard about the nature and economics of the Regent's Park

---

30 On these matters see Grigson, *Menagerie*; Ito, *London Zoo and the Victorians*; Jackson, *Menageries in Britain 1100–2000*; Kisling, *Zoo and Aquarium History*; Loisel, *Histoire des Ménageries de l'Antiquité à Nos Jours*; Simons, *The Tiger That Swallowed the Boy*; J.L. Middlemiss, *A Zoo on Wheels: Bostock and Wombwell's Menagerie* (Burton on Trent: Dalebrook Publications, 1987) and K. Scrivens and S. Smith, *Manders Shows and Menageries* (Newcastle under Lyme: The Fairground Society, no date). The other important sites for animal display were the private menageries, large and small, many of which were attached to country estates. The two most important of these were Lord Derby's menagerie at Knowsley Hall near Liverpool and Walter Rothschild's collection at Tring. Both Lord Derby and Lord Rothschild were important friends to the Zoological Society and provided many animals for Regent's Park. See C. Fisher (ed.), *A Passion for Natural History: The Life and Legacy of the 13th Earl of Derby* (Liverpool: National Museums and Galleries on Merseyside, 2002); and M. Rothschild, *Walter Rothschild: The Man, the Museum, and the Menagerie* (London: Natural History Museum, 2008).

Gardens. By the late 1840s, it was clear that the exclusivity of access which was a foundational principle was not going to deliver a long-term business, especially as the zoo had increasingly to compete with new institutions, particularly in Germany and Holland, for the rare animals that created waves of public interest as well as furthering the scientific mission.[31] The Jardin des Plantes in Paris in particular loomed large on the Zoological Society's horizons, and the aim to have a better zoo than the French is never far away, implicitly or explicitly, from the ideological constructs that underpin the foundation and development of the Zoological Gardens. On 14 June 1851 the *Illustrated London News* observed, in one of its periodic reports on Obaysch, that:

> A self-supporting Society, subject, consequently to every possible fluctuation to which bad management or the caprice of the public taste can give rise, has not only succeeded in raising itself to the character of a truly national Institution, but has succeeded in effecting much more than the cognate Institutions of other countries, supported by their respective Governments ... Such, however, is the self-reliance of the Anglo-Saxon race, that its undertakings and resources seem to flourish most when left most to their own resources, and most independent of the trammels of official machinery.

This magnificent statement of liberal principles ignored the fact that the Jardin des Plantes, which operated on very different lines and was a state institution, did just as well or better as the Zoological

---

31 In fact, as recently as 1991 a decision was taken to close Regent's Park Zoo and to concentrate the live animal side of the Zoological Society's work at Whipsnade, where costs were significantly lower. Predictably this caused a national outcry and a combination of interventions enabled the zoo to develop pathways back to sustainability. Nevertheless, the fact that an historic national institution could so recently have all but closed offers a view into the precariousness and expense of the zoo business and another reminder of the importance of Obaysch in saving the zoo before.

Introduction: Why Obaysch?

Gardens in many key respects, while the Artis Zoo in Amsterdam operated on very similar lines to the London Zoo and had only that year moved away from an exclusive membership-based access model to a limited open access system where, in September of each year, anyone could buy a reasonably priced ticket.[32]

Nevertheless, Britain still had several advantages. In particular its imperial dominance in India and Australasia gave it exclusive access to a huge range of creatures. Before the construction of the Suez Canal most ships coming in from Asia and Africa made London their first port of call. This enabled British dealers and institutions to have first pick of any animals that happened to be on board. Ships coming from the Americas called first at Liverpool and this likewise gave the British trade an advantage. But these advantages were time-limited and, as we have seen, not necessarily guarantees of exclusive rights. Obaysch and Adhela may have come to England as part of a diplomatic deal with the authorities in Egypt but the Egyptians were equally concerned to keep the French on side, and so Coco and Bichette arrived in France only a few years later as the new stars of the Jardin des Plantes.

It was clear to the Zoological Society that to sustain the Zoological Gardens it would need to allow more access to the general public, but this raised a number of alarming questions. How would people behave? What days and for how long should the gardens be generally open? What should the price be? Would more general access compromise the society's scientific mission and turn the Zoological Gardens into a place of mere popular entertainment? On one side of the debate stood an alliance of old-fashioned snobs who didn't want working-class people in their zoo, and Sabbatarians who believed that Sunday opening was irreligious (most working-class people only really had Sundays free, although the observance of 'Saint Monday' was still surprisingly common in the

---

32 See D. Mehos, *Science and Culture for Members Only: The Amsterdam Zoo Artis in the Nineteenth Century* (Amsterdam: Amsterdam University Press, 2005).

early Victorian period). On the other side stood pragmatists who knew the zoo had to make more money or go bust and idealists who genuinely believed that the society's scientific and educational mission could only be enhanced if its benefits were made more or less universally available. The economic facts were these: between 1836 and 1848 annual revenue decreased from £19,123 to £8165. In 1843 the society ran a deficit of £3000, meaning that by 1848 its capital reserve stood at a mere £3836. This was not enough to pursue capital projects, nor was it enough to provide the investment fund necessary to acquire new animals.[33]

The zoo had been founded on the basis of a £3 joining fee and a £2 annual subscription for fellows, who then enjoyed access to the gardens. They could bring a friend (who had to be accompanied by them on their visit) for a shilling per day.[34] This system was at the bottom of the increasingly perilous state of the society's revenue streams. The problem was solved by David Mitchell, who, in 1847, was appointed secretary (what we might now call chief executive officer) to the Zoological Society; in that year the zoo finally opened its doors and offered entry to the general public, without an accompanying fellow, on Mondays and Tuesdays as well as over Easter and Whitsun for a shilling. The next year this was extended to all week days (with Saturdays and Sundays being reserved for private access of the fellows). At the same time children were allowed to come in for sixpence and, on Mondays, adult visitors were also charged only sixpence. These prices were still high, but they compare not unfavourably with those charged by the menageries and private zoos. Sunday access was still a problem. Fellows could come in on a Sunday, so clearly Sabbatarian sentiment extended only to the management of working-class leisure. And the

---

33 See Ito, *London Zoo and the Victorians*, pp. 81–137, for a full exploration of these issues.
34 A fellowship subscription now costs £141 and entry to the zoo for the day £29.75. These figures represent considerably lower real values than those obtained when the zoo was founded.

Introduction: Why Obaysch?

feared poor behaviour never eventuated. In any case, the records of the society show that some of the fellows were capable of pretty bad behaviour in spite of their money or titles. It was not easy for a humble zookeeper to get a lord not to smoke a cigar in a non-smoking area or a bohemian society artist like Rossetti not to poke a wombat with his cane.[35] But Mitchell had caught the *Zeitgeist* and recognised that institutions like the Zoological Gardens had a social mission to play in a society where change was in the air – and that institutions which were not part of that change would eventually succumb.

It is important to note that one function of zoos at the time was 'acclimatisation'. This was the project to introduce new animals into Europe to supplement the agricultural industries and, especially, to solve the problem of producing sufficient meat-based protein for a growing working class that needed better food. This project was a part of the operational principle of the Zoological Gardens, especially under Mitchell, with Buckland as an enthusiastic supporter, but it was also important to the European zoos. In the colonies, Melbourne Zoo and Adelaide Zoo were founded essentially as a laboratory for acclimatisation, and similar experiments were conducted in New Zealand (where the impact of acclimatisation on a fragile eco-system which had no native mammals is still an ongoing problem to be managed).[36] The *Belfast*

---

35 G. Vever's *London's Zoo* (London: Bodley Head, 1976), pp. 54–65, gives some amusing examples of badly behaved visitors.
36 See C. de Courcy, 'Evolution of a Zoo: A History of the Melbourne Zoological Gardens, 1857–1900' (unpublished MA thesis, University of Melbourne, 1991); C. de Courcy, *The Zoo Story: The Animals, the History, the People* (Melbourne: Penguin, 1995); C. Lever, *They Dined on Eland: The Story of the Acclimatisation Societies* (London: Quiller Press, 1992); R. McDowall, *Gamekeepers for the Nation: The Story of New Zealand's Acclimatisation Societies, 1861–1990* (Christchurch: University of Canterbury Press, 1994); I. Parsonson, *The Australian Ark: A History of Domesticated Animals* (Collingwood, VIC: CSIRO, 2000); and E. Rolls, *They All Ran Wild: The Story of Pests on the Land in Australia* (London: Angus & Robertson, 1984).

Newsletter (15 January 1871) added, as an afterthought to a piece on the death of Adhela's baby speculated, perhaps humourously, it is hard to tell, on the economic opportunities of acclimatisation and the large amounts to be made, potentially, from hippos. David Mitchell later went to run the Jardin d'Acclimatation in Paris in 1859, but died there soon after he took up his appointment. The fact that this French zoo wanted his services speaks to the respect with which his skills were internationally regarded. Although there was no plan to acclimatise hippos in nineteenth-century Europe, there was such a suggestion in early twentieth-century America, where it was proposed in Congress that stocking the Mississippi with hippos would add a useful supply of meat to a nation wracked by the Depression.[37]

This liberalisation of access began to build the income that would create a sustainable business, but the problem still remained of how to develop repeat custom, as there was no point in having a few boom years followed by decline when everyone had seen everything there was to be seen and didn't come back. Once again David Mitchell solved the problem. He understood the fundamental problem of a tired experience so set about strategically developing the 'charms of novelty', partly by a major expansion of the collection and more attractive display but also through the 'star' animals system. Obaysch was the first such star animal and, by 1852, the society's income has bounced up to £26,600 with a surplus of £4000 which Mitchell insisted be made available for investment.

The star system soon became very explicitly part of the zoo's public image and on 24 December 1853 *Punch* ran an interesting article entitled 'The Fashionable Zoological Star', which was stimulated by the arrival of the American ant-eater:

---

37 On this see J. Mooallem, 'American Hippopotamus', *Atavist Magazine* 32 (December 2013), https://bit.ly/2iVT7lo.

Introduction: Why Obaysch?

We are sorry to see that the Zoological Gardens have lately got into the 'Star system.' Not content with a good working company of bears and monkeys, they must have particular 'Stars' to bring the million in. Some time ago it was a hippopotamus, who made all London run after him. Then there was the baby elephant, who was a source of great interest to mothers. After them followed a chimpanzee, and a serpent-charmer, and a whole forest-full of humming birds, and we cannot recollect what else. All of them, however, were great attractions in their way; in fact, it may be said that the animals have lately taken the shine out of the actors. As the theatres have gradually become more empty, the Zoological Gardens have perceptibly become more crowded. What actor, recently, has had anything like the success that for a whole season ran panting, pushing and squeezing after the Hippopotamus? It was a fight of parasols to get near him – it was joy greater than that of a new gown to have seen him! What is the reason for this strange preference? Is it because the public prefers Nature to Art? – or is it because the actors speak, and the animals do not?

This recaps the astonishing popularity that Obaysch enjoyed and the crush around his pen, but it also adds two interesting side notes. The first is that going to the zoo may be a gendered experience – note the parasols and gowns. The second is that the zoo takes its place as one of several options for middle-class entertainment and its adoption of the star system explicitly recognises that. The remainder of the article consists of some good jokes about the ant-eater and the point that animal stars unlike human stars don't ruin their managers with excessive salaries, are content with regular meals and don't go on strike 'for more beans'. The piece shows how well Mitchell had laid down his strategy and also how that strategy was not just about star animals but also about the development of what we now call a more inclusive appeal for his product and a positioning of that product within the leisure market.

The addition of new attractions continued to build the business. In 1854, for example, the new 'Fish House' was 'bringing in more money than did the Hippopotamus.'[38] As late as December 1866, *Fun* magazine reflected the importance of Obaysch and subsequent hippos to the finances and reputation of the zoo:

> We understand that at the dinners of the Zoological Society it is usual for the toastmaster, when a health is drunk 'with the honours,' to lead off the cheers with 'Hip-hip-hippopotamus-hurray!'[39]

Fifteen years after public interest in Obaysch was at its height, this snippet shows just how significant the hippos' contribution to the development of a sustainable business model for the Zoological Society was commonly perceived to be, even after the initial excitement surrounding Obaysch had subsided. And up until the late twentieth century, when the threat of closure again loomed and necessitated another major change in the business model, the London Zoological Gardens remained one of the primary sites of zoological inquiry and display in the world, a place to visit many times and a beacon of imperial achievement. This may well not have happened without Obaysch, as his entry into the zoo not only provided the society with a massive injection of cash at the point when it was most needed but also, and more importantly, reset the zoo as a place of great attraction where learning and entertainment could be enjoyed in equal measure.

Let us now look at his life in detail.

---

38 Ito, *London Zoo and the Victorians*, p.120
39 *Fun*, 15 December 1866, p. 142.

1
The Life and Times of Obaysch the Hippopotamus

Obaysch's life story has two parts: the first part is very much shorter than the second. For up to a year or so he lived with his mother on the Nile as a wild hippopotamus on or near the island of Obaysch. For a further twenty-seven years he lived, sometimes alone and sometimes with or near other hippos, in a pen in the Regent's Park Zoological Gardens. That period has been the exclusive focus of everyone who has written about him before. But he had a life before that time, albeit a brief one, and it deserves attention if we are truly to understand what his long captivity meant.

He was born almost certainly in 1848 but possibly in 1849. For reasons related to the life cycle of the hippopotamus and the fact that in May 1850 he was referred to as eighteen months old, I favour 1848. I also do not believe a hippo as young as Obaysch would have been when he was captured if he had been born in 1849 could have survived the arduous journey from his home range back up the Nile. Having said that, the naturalists at the time were all convinced he was very young and a hippo captured by John Petherick in 1858 allegedly still had the umbilical cord attached when taken and survived not only to be transported down the Nile but also then to come to England. So with the requisite care it may have been possible to preserve the life of an infant hippo in transit. This allows for a birthdate in early 1849 for Obaysch if we discount the notion that he was thought to be eighteen months old in May 1850. If we accept the estimate of eighteen months that would give him a birthday in November 1848 which feels right.

But the late 1840s and particularly 1848 were momentous years and across Europe. Revolutionary and nationalist movements were manifesting themselves in all manner of outbreaks collectively forming what is occasionally referred to as The People's Spring. There was the beginning of the anti-Bourbon revolt in Sicily which would eventually lead to the unification of Italy. Louis Napoleon (who subsequently became Napoleon III) took control of the revolutionary unrest in France and proclaimed the Second Republic. Conflicts broke out all over the German states and there were barricades on the streets of Berlin (which did not prevent a sea lion being exhibited there in a barrel at the same time – anyone who has been in a city on the day a revolution or a civil war breaks out will know that these things can be surprisingly localised and that just round the corner from the tear gas and placards life goes on as usual, sea lions and all).[1] The monarchy's powers were trimmed in Denmark and nationalistic fervor in the duchies of Schleswig and Holstein would lead to a war with Prussia which the Danish won but which set up a chain of events that, arguably, would not lose momentum until the Berlin Wall was torn down in 1990. The polyglot and multicultural Austrian Empire was likewise riven with conflict, especially in Hungary. Sweden, Poland, Belgium and even Switzerland experienced unrest (in 1847 the Swiss had the most civilised internal war it was possible to have with generals on both sides having it as their avowed intent to kill as few of the enemy soldiers as possible). Marx and Engels' *Communist Manifesto* was published. In Egypt the effects of all this were not felt in social

---

1   E. Ames, *Carl Hagenbeck's Empire of Entertainments* (Seattle: University of Washington Press, 2009), pp. 9–10.

1 The Life and Times of Obaysch the Hippopotamus

unrest so much as in the shifting relationship between the Ottoman Khedivate and the British. Britain was making its influence increasingly visible in the region and the country had developed a complex relationship with Constantinople. This was largely motivated by a desire to shore up the Turks as a bulwark against the Russians who might, in the worst Imperial nightmare, one day come streaming down through Afghanistan and the Khyber Pass to appear, moustaches bristling, at the gates of Delhi. The Great Game had not yet properly started but the chess pieces were all in place and Obaysch can, arguably, be seen as one of them.

The year 1849 was an important one for the European relationship with hippos for another reason. In 1843 the American scientist Samuel G. Morton, who was at that point studying in Liberia, had noticed hippopotamus-like animals which were referred to locally as water cows and had been thought of as a kind of wild pig by previous European explorers. When he saw the skulls he concluded, correctly, that they were in fact a kind of hippopotamus and initially categorised them as simply a small-sized sub species of the common African hippo (*Hippopotamus minor*). But, by 1849, he had seen enough specimens to realise that they were in fact a different animal of an independent species and categorised them as such in an article published in the *Journal of the Academy of Natural Sciences of Philadelphia* which followed up on an account of his preliminary findings published in the same journal in 1844.[2]

---

2   S.G. Morton, 'On a Supposed New Species of Hippopotamus', *Journal of the Academy of Natural Sciences of Philadelphia* 2 (1844), pp. 14–17, and 'Additional Observations on a New Living Species of Hippopotamus', *Journal of the Academy of Natural Sciences of Philadelphia (Second Series)* 1 (1849), pp. 231–239. The endangered pygmy hippo is such a rare and reclusive animal that it was, amazingly, not photographed in the wild until 2004. A recent doctoral thesis offers an excellent overview of the pygmy hippo (G.L. Flacke, 'The Pygmy Hippopotamus: An Enigmatic Oxymoron', University of Western Australia, 2017). At least two subspecies are known to have become extinct in recent times. A Nigerian subspecies (*Choeropsis liberiensis heslopi*) may have finally disappeared as little as seventy years ago without ever being

A pygmy hippo didn't arrive in Britain until 1873 when one was transported to Dublin Zoo (alas, it never recovered from the trauma of the voyage and died almost immediately) but the fact was that people now knew that *Choeropsis liberiensis* was out there even if they couldn't see one in the flesh.[3]

For a number of years the Zoological Society of London had wished to acquire a hippopotamus. In 1847, an expedition to capture one was planned but abandoned when a suitable guide could not be found. But now Sir Charles Adolphus Murray, the British consul-general in Cairo, and ever afterwards known, apparently to his great delight, as 'Hippopotamus Murray', was asked to see if it might be possible to persuade the Egyptian authorities to capture one which could then be shipped to London.[4] It was a fortuitous moment for such a request as the government of Egypt was in flux and the long-reigning Muhammad Ali and his short-lived son Ibrahim had just been replaced by Pasha Abbas I. His relationship with Britain was a difficult one as he was neither a conservative nor a modernizer – the two opposite, but handily facile, categories into which the British tended and still tend to place Arab rulers – and so it was not easy to establish a common ground on the basis of a clear program. However,

---

studied as a live animal, while the Madagascan subspecies (*Choeropsis madagascariensis*) disappeared sometime in the last 500 years. Australia has, alas, never had hippos, although the megafaunal *Zygomaturus trilobus* is occasionally referred to as a marsupial hippo because of its probably amphibious habit. But it was not a hippo and is distantly related to wombats.

3 C. de Courcy, *Dublin Zoo: An Illustrated History* (Wilton, Cork: Collins, 2009), pp. 54–55. This hippo was sent from Sierra Leone and was a donation to the zoo, as were many of its animals in the early days. It can still be seen, stuffed, in the Zoological Museum of Dublin's Trinity College. Diana, the first pygmy hippo at Regent's Park, arrived in 1913.

4 Murray is an interesting and engaging character. See H. Maxwell, *The Honorable Charles Murray KCB: A Memoir* (Edinburgh: W. Blackwood, 1898); P.C. Peck, 'Hippopotamus Murray', *Scotland Magazine* 83 (2015), p. 56; and D. Weinding and M. Colloms, 'Hippopotamus Murray', *West Hampstead Life* (March 2014). A wonderful daguerreotype taken by Antoine Claudet in 1851, the year after Obaysch had left his care, shows Murray in splendid tartan trousers reclining with his hookah while two Egyptian servants stand ready with trays.

1 The Life and Times of Obaysch the Hippopotamus

like other pashas before him he was seriously interested in horses, especially in the breeding of Arabians and everything that went with this. It seems, therefore, that it was possible for Murray to offer some personal rewards in the form of greyhounds and deerhounds together with the loan of a trainer in exchange for the hippo and some other animals including a leopard and a cheetah. As noted in the previous chapter, hippos had enjoyed a somewhat protected status in Egypt; or at least, means of peaceful co-existence between hippos and farmers had been established. But, by the middle of the nineteenth century, things had changed, to the detriment of the hippos. More modern methods of farming, especially changes to irrigation systems, had put habitat under threat (this ongoing process remains the greatest single threat to wild hippos, much greater than poaching or hunting for 'bush meat'). In addition, the government in Cairo was increasingly unable to control travel down the Nile. This led to an upsurge in hunting by Europeans. All this contributed to a significant downturn in hippopotamus numbers, to the point where they became a rarity: in 1878, when the Khedive wanted a hippopotamus for his own private zoo one had to be imported into Egypt from another part of Africa.[5] By 1892 an article in the *Graphic* (10 September) sadly concluded that:

> There can be no doubt about the rapid destruction of these animals ... The influx of Europeans and the civilization of Africa will, in a few years, exterminate this powerful animal, which is not likely to be domesticated or rendered serviceable to mankind.

So Obaysch was born at a time when his previously common and unofficially protected species was coming under threat and in the beginnings of decline. There were fewer places for hippos to live and the risk of death at the hands of hunters grew daily. At the same time the changing relationships between the Egyptian and the British authorities and the personal interests of Pasha Abbas (not least, it has

---

5   Mikhail, *The Animal in Ottoman Egypt*, p. 171.

been suggested, his desire to surpass his predecessor Ibrahim, who in 1849, at the very end of his reign, had given several potentially mating pairs of animals – two giraffes, two dromedaries, two Arabian oryxes, two addax antelopes, two ostriches and two gazelles – to the Zoological Society) created a diplomatic space within which it was possible to agree to the capture and exportation to London of a wild hippopotamus. It was part of a deal that was partly international relations, partly the mutual projection of soft power, and partly a bribe.[6] Hippos were not there simply to be given away, to be captured without permission or bought for cash. According to the *Illustrated London News* (1 June 1850) there was also at this time an American based in Alexandria offering the massive sum of £5000 for a live specimen. If this is true, and I think it (or something near to it) probably is, it is significant that the Pasha preferred the development of a diplomatic relationship with the British over a substantial and handy injection of cash into his personal fortune. This shows what a significant animal Obaysch was to become.

Soldiers were dispatched downstream with orders to bring back a hippo. As they left, Obaysch would probably have just about been born.

In his first year Obaysch would have lived mainly in the river. He would have been born under water. He would then have had to make his way to the surface under his own steam if he wanted to take his first breath. Hippos are suckled for much of their first year, with weaning beginning at about eight months. We should therefore assume that Obaysch was old enough to have been weaned – or almost completely weaned – when his mother was killed, although the vast amounts of milk he was given to drink on his journey to London suggest that his handlers may have thought otherwise. Obaysch would also have been given his mother's faeces to eat.

---

6 Rare animals have frequently been used as diplomatic tokens in soft-power games, and continue to be. See J. Simons, 'The Soft Power of Elephants', in N. Chitty, L. Ji, G. Rawnsley and C. Hayden (eds), *The Routledge Handbook of Soft Power* (London: Routledge, 2017), pp. 177–184.

Hippos are born with sterile intestines and need to establish the beneficial gut bacteria that will enable them to digest the grasses on which they will graze (often on well-defined riparian 'hippo lawns'), on which they will depend for most of their nourishment for the rest of their lives.

After eight months in the womb, Obaysch would have weighed about eight stones (50 kilograms) at birth and been about four feet (122 centimetres) long. He would have lived on his mother's back at first, but as he grew older he would have spent more time in the deeper water and would even have fallen asleep on the bottom of the river. Baby hippos need to breathe about every two minutes or so, but this is an automatic process: when they are resting they rise and fall, like hippo-shaped balloons, or like the whales to which they are distantly related, floating from the river bed to the surface, so that they don't drown in their sleep. During the day he would rarely have emerged from the water; when he did, he would have exuded a red antibiotic fluid (known as 'bloodsweat' although it is neither sweat nor blood) that protected his skin from the damaging effects of ultra-violet light as effectively as a high-factor sunscreen. One observer of Obaysch pointed out that, thanks to bloodsweat, 'You can spoil a new pair of gloves, indeed, by just patting a hippo once.'[7]

Hippos have no ability to sweat and as their skin has only its bloodsweat for protection they are, perforce, nocturnal animals. They spend as much as sixteen hours of each day asleep.

Being a young male hippo is a tricky business. Obaysch would have had to learn quickly how to negotiate the territorial boundaries of adult males. Although hippos like to cluster together they are not very social, and Obaysch would not have formed an especially close bond with his mother. Hippo mothers and daughters, however, do appear to form a social relationship, and this may be important to our understanding of Obaysch's later life. His mother was probably

---

7   A.T. Elwes and T. Wood, 'The Zoo Past and Present', in *Sunday Reading for the Young* (New York: E. & J.B. Young & Co., 1903).

one of ten or more females in the bloat of an alpha male, who would rule a territory of river about 275 yards (250 metres) in length. 'Bachelor' male hippos would also have been part of the bloat and the alpha male would tolerate them as he would have tolerated Obaysch, provided that they behaved with appropriate submissiveness. A powerful male might control a bloat of 100 animals, and a bloat of 200 is possible if rare. But given what we know about hippo numbers in Egypt in the 1840s and the nature of Obaysch's capture, it seems unlikely that he and his mother were members of such a large group. Obaysch's mother (and baby Obaysch) would have spent their days mainly with the other females of the bloat while the bachelors crowded together elsewhere and the alpha male rested alone grumbling to himself. We do not know anything about Obaysch's mother except that she must have been at least five years old to have had a calf. His father, about whom we know nothing, as only his mother appears in the brief drama of his capture, would have been at least seven years old.

That Obaysch survived his first year shows that his bloat was in reasonably good shape. Infanticide by alpha males tends only to happen under stressful conditions and it seems that his mother had not had to defend him. Male hippos do fight each other for dominance on occasion and do each other fearful damage but, by and large, they are not threatened by other animals. Babies can be taken by crocodiles and there is some evidence that lions, acting as a pride, may attack an isolated hippo. Hippos frequently attack crocodiles to drive them out of their territory and are not prepared to share their stretch of river with them. They also attack humans without provocation and are still one of the most dangerous of African animals. They kill an estimated 3000 people per year. However, although Obaysch may well have lived in a healthy bloat, the fact that the hunters were so easily able to kill his mother and separate him from the rest of the hippos suggests that it was quite a small bloat, which would be consistent with what we know about

1 The Life and Times of Obaysch the Hippopotamus

hippopotamus numbers on the White Nile in the middle of the nineteenth century.

In his first year Obaysch lived in a relatively tranquil world. He suckled and was weaned onto grass, gradually building up his intake to the 75 to 110 pounds (35 to 50 kilograms) he would eat every day by the time he was fully mature at seven. He might have seen conflict between male hippos and maybe even an attack on a crocodile. If the crocodile tried to escape by land an enraged adult might have chased it at a speed of up to twenty-three kilometres per hour before biting it in two with his vicious ivory teeth.[8] He would have seen dung flying through the air as the alpha male marked his territory by defecating and spinning his tail at the same time and 'yawning' at bachelor males who were not being sufficiently deferential as they passed him. He would have heard various wheezes, grunts and snorts of up to 115 decibels, including the unique hippopotamus vocalisation, which travels through air and water at the same time. At the end of his first year Obaysch would have been learning the skills of being a wild male hippo and could have been expected to live, barring mishaps or death by wounds sustained in a fight with another male, until he was about forty or fifty years old (not the twenty-seven or twenty-eight years he managed in captivity), but he would also have expected to stay under his mother's protection until he was at least three years old. And then the hunters came.

Male hippos, unlike female hippos, continue to grow throughout their life. A mature male can reach five tons in weight. Once females reach a ton and a half they rarely grow more. The massive scale

---

8   Hippopotamus ivory was widely used in the manufacture of false teeth in the middle of the nineteenth century – just when the post-1815 'Waterloo teeth', culled from the dead (and some say the nearly dead) of that great battle, were wearing out. But by the latter part of the century the material was also being criticised as too easily broken and having a tendency to go yellow, giving a less than pleasing cosmetic effect. Perhaps the most famous hippopotamus teeth were the decidedly ill-fitting set sported by George Washington. The Duke of Wellington himself had false teeth which were 'Waterloo teeth' at the front and gold and walrus ivory at the back.

of a hippo's body needs to be understood as we contemplate any project to keep one captive in the middle of a city. The relative sizes of the mature male and female hippo body will also play a part in understanding and interpreting at least one incident in Obaysch's captive life.

In early August 1849, Obaysch had been hidden in some bushes on the Nilotic island of Obaysch (sometimes transliterated as Fobaysch), some 1400 miles (2250 kilometres) from Cairo when Abbas's hunters surprised his mother.

The commonly received version of his capture tells us that the wounded animal rushed to try and save her calf but succeeded only in giving his concealment away. The hunters killed the mother and grabbed him but his skin, made slippery by the red bloodsweat, enabled him to wriggle free and make for the water. If he could sink to the bottom he might escape. But one of them pierced him with a hook. This caused a scar on his back that was visible until the day he died, although as he grew bigger the scar's relative position on his body changed until it was nearer to his back leg than to his middle. David Mitchell's more or less contemporary account of his capture casts a rather heroic and romantic light on the whole affair and transports the reader back to the time of the Pharoahs:

> The hunter, however, with the presence of mind which characterizes a good sportsman, seized his spear, and, with the sharp side-hook, which has been in fashion in Egypt for three thousand years, he succeeded in arresting the headlong plunge of his prize, without inflicting greater damage than a skin-wound, which is marked by a scar on his ribs to this day.[9]

---

9   This is the account given by D.W. Mitchell in *A Popular Guide to the Gardens of the Zoological Society of London* (London: Zoological Society, 1852), p. 54. Interestingly the account given by 'Æthelwode' in the *Sporting Magazine* 46 (1897), pp. 386–391, makes no mention of Obaysch's mother and places the capture in July 1849. This is an especially interesting account as it suggests

1 The Life and Times of Obaysch the Hippopotamus

Obaysch was hauled into a boat and struggled until it capsized. But pain and fear had exhausted him and his captors were, eventually, able to secure him ready for transport.

However, I think that the realities of Obaysch's capture were somewhat different and that the story of the hook is that something that crept in, either as a story told by the hunters themselves to make their procedures look less brutal or by Mitchell and others to play down the naked acquisitiveness that led to Obaysch's capture. Eduard Rüppell was a German naturalist and explorer who, in 1829 published his *Reisen in Nubien, Kordofan und dem peträischen Arabien*. This work included an account of a hippopotamus hunt near Dongola, which was in the same region as the island from which Obaysch was taken. This account was widely translated in the popular periodical press in Britain from the early 1830s onwards. This extract is from the *Penny Magazine* for 2 June 1832:

> The harpoon with which the natives attack the hippopotamus, terminates in a flat oval-shaped piece of iron, three-fourths of the outer rim of which are sharpened to a very fine edge. To the upper part of this iron one end of a long stout cord is attached, and the other is tied to thick piece of light wood ... When the huntsman is about seven paces from the beat he throws the spear with all his might, and if he is a good marksman the iron pierces through the thick hide, burying itself in the flesh deeper than the barbed point. The animal generally plunges in the water; and though the shaft of the harpoon may be broken, the piece of wood that is attached to the iron floats on the surface and shows what direction he takes ... The huntsmen now pull the rope, when the monster, irritated by the pain, seizes the boat with his teeth and sometimes succeeds in crushing or capsizing it ...

---

that the pseudonymous author had an informant – had he met Murray? – who furnished him with details not found in other sources.

The hippopotamus was then speared and bludgeoned to death by those of the huntsman's four or five companions who had managed to survive the hippo's initial attack on their boat.

All of the elements of Obaysch's capture are, on careful examination, to be found here. The scar on his side is not from a hook used on the spur of the moment but from the barbed blade of a traditional hippopotamus-hunting spear being used in the traditional way. The scar is easily visible in the photograph of Obaysch on the front of this book. It looks like a puncture wound, which would further bear out the idea of a traditional spearing rather than a glancing attachment of a hook. He wasn't strong enough to bite through the hunters' canoe but he did apparently capsize it. They said that this happened once he was inside, but it seems equally likely that it occurred when they were reeling him in on their harpoon line. Of course, we cannot know the truth, but we do know that Egyptians and Sudanese had been capturing or killing hippos in the fashion witnessed and recorded by the intrepid Dr Rüppell for many hundreds of years. It seems to me likely that when the soldiers of the Pasha's expedition reached Dongola, they cast around for some local men who knew how to hunt hippos, found some, and explained to them that they wanted one captured, not killed. These men would have known that their traditional methods could well enable them to obtain a baby hippo and they may or may not have kept quiet the inevitable truth that it would be slightly damaged in the process. It would then have been an easy matter to explain Obaysch's wound as a last ditch attempt to keep hold of their precious prize rather than as an essential part of the process. No doubt the lieutenant in charge was only too pleased to have what looked like a viable hippo and all too keen to get back to Cairo with few questions asked after an arduous time in the desert.

And so the story of the slippery little hippo sliding from the hunter's grasp and being pulled in by a hook at the very last moment became a fact of Obaysch's life and a key element in his keepers' narrative.

1 The Life and Times of Obaysch the Hippopotamus

There now stretched a long journey ahead, as having captured their hippo the hunters had to transport him to Cairo. The fact that his kidnap had been commissioned by Pasha Abbas himself meant that careful arrangements were already in place to ensure his safe conduct. The history of European expeditions into Africa to capture wild animals is littered with the dead bodies of the animals. Although the profits to be made from the trade made the wastage acceptable, if not exactly desirable, from a business perspective.[10] Abbas had clearly invested a lot of 'face' in this particular animal and did not intend to have it die on him. So he provided a boat which had been custom-built specifically for hippo transportation together with an escort of ten Nubian soldiers and their lieutenant. We know little about the journey but clearly the diet of cows' milk and dates – probably the soldiers requisitioned milk and maybe even cows or goats from the villages they passed – kept him in good shape. By 14 November 1849, between five and six months after he was captured, Sir Charles Murray was able to write to the Zoological Society to inform them that he had taken delivery of the animal, that it had been escorted from the Pasha's palace by his chief officer, and that it was now living in the yard at the back of his house. Murray reported that 'It is only five or six months old [this surely cannot be right, as it would mean Obaysch had been only a few weeks or even days old when captured], and still lives entirely on milk [this could be true even if he was almost a year old]; I think a fresh importation of cows will be necessary in Cairo, as our little monster takes about thirty quarts of milk a day for his share already.'

---

10 See N. Rothfels, 'Catching Animals', in M. Henniger-Voss (ed.), *Animals in Human History* (Rochester, NY: University of Rochester Press, 2002), pp. 182–228; and J. Simons, 'The Scramble for Elephants' in M. Boyde (ed.), *Captured: The Animal within Culture* (Houndmills, UK: Palgrave Macmillan, 2014), pp. 26–42.

On the same day we have a slightly different eye witness report written by Samuel Shepheard, the proprietor of the famous Shepheard's Hotel and an intimate friend of the Pasha:

> In the spring of the year, Inshallah! Your London visitors will be gratified with a sight of one of our river monsters – the Hippopotamus. Our consul here, the Honble Mr Murray, having expressed a wish to have one sent to England, where such a thing had never yet been seen, His Highness ordered off a Captain with a troop of twenty men to proceed up the interior of Africa to follow the course of the white river and not to return without one. After having been five months away they have returned and brought a very fine specimen 5 or 6 months old. He is yet suckling and drinks 80 pints of milk at a meal and, unwieldy though he is, plays about with the keeper like a great fat pet pig with a square head. The Pasha has presented him to Mr Murray.[11]

Just as Pasha Abbas had ensured that Obaysch's transportation was as safe and secure as could be arranged, so Murray had set up suitable quarters for Obaysch. The yard had a bath which could be gently warmed if necessary during the Cairo winter and he had also secured the services of the experienced Hamet Safi Cannana (his name is transliterated in various ways in the newspapers and other sources, and he was also known as Hippopotamus Johnny, but for the remainder of this work I will refer to him as Hamet) as Obaysch's keeper. In one source Hamet is noted as having been 'engaged at Cairo from his skill and experience at the management of animals' but I have not been able to determine in what it consisted or where he acquired this.[12] Obaysch spent the next six months in

---

11 Quoted on the Zoological Society of London's artefact of the month blog in the context of Gawen's statuette of Obaysch (discussed below), which was commissioned by Shepheard; http://bit.ly/ZSLObaysch.

12 H.D. Northrop, *Earth, Sky and Sea* (San Francisco: The J. Dewing Co., 1887), p. 243.

1 The Life and Times of Obaysch the Hippopotamus

the temporary quarters in Murray's yard and, on the evidence of Murray's letters back to the Zoological Society he thrived:

> The Hippopotamus is quite well, and the delight of every one who sees him. He is tame and playful as a Newfoundland puppy; knows his keepers, and follows them all over the courtyard; in short, if he continues gentle and intelligent as he promises to be, he will be the most attractive object ever seen in our Garden, and may be taught all the tricks usually performed by the elephant.[13]

Obaysch certainly delighted everyone who saw him, but Murray's vision of a performing hippo thankfully never came into being at least in his case. Although, as we shall see, there were certainly performative elements in the way Obaysch was exhibited. Hamet proved to be a faithful friend to Obaysch in his subsequent captivity and was obviously an effective manager of his diet and health. Dr Henry Abbott, an English doctor and long-term resident of Cairo, where he had amassed an important collection of antiquities that was one of the attractions of the city to tourists, also helped to look after him. Although Abbott was a medical practitioner and so would have had a general knowledge of mammalian physiology, it is unlikely that he knew anything like as much about hippos as Hamet. Dr Abbott is mentioned as one of Obaysch's keepers by an English tourist who was in Cairo in February 1851 and went to see the hippo as one of the numerous 'living and dead curiosities' that were on hand to beguile away the time of the inquisitive visitor to the city.[14]

---

13 Quoted in D. Manley, 'A Traveller from Egypt', *Bulletin of the Association for the Study of Travel in Egypt and the Near East: Notes and Queries* 21 (2004), pp. 21–22. See the commentary on Bucheet below – at least one performing hippo did appear in London during Murray's lifetime.

14 'A Hippopotamus for the Zoological Society', *Illustrated London News*, 25 May 1850. Dr Abbott had an interesting life post-Obaysch. In 1853 he took his collection to New York and set up the Egyptian Museum, sometimes referred to as the Egyptian Gallery, on Broadway. He became a friend of Walt Whitman, who was deeply fascinated by ancient Egypt. The collection – a tenuous link

At this time Obaysch was clearly recovering from the ordeal of his capture and transportation and one English newspaper reported that he had 'considerably improved in dimension and appearance under the kind care bestowed on it by its European keepers.'[15] Why a European who would have had no idea what a hippopotamus might need or want would do better than Hamet, who seems to have known the animals well, is unclear. However, the racialised commentary on Obaysch's life in Egyptian captivity prior to his export to London casts an interesting light on Victorian attitudes to animals. The kindness of Europeans is implicitly contrasted with the brutality of the Africans who had taken him from the wild, albeit at the behest of those very same 'kind' Europeans. And, as we shall later see, Hamet was frequently racially vilified during his residence at the Zoological Gardens.

During Obaysch's stay with Murray, Pasha Abbas was so pleased with the positive response to his gift, and aware of the diplomatic and personal advantages that might accrue by an even greater effort, he sent off another expedition. This one had the mission of bringing back a female hippo so that he could present the Zoological Society with a pair just as his father Ibrahim had done in 1849. However, the soldiers this time returned empty handed, which shows again how difficult it was to secure a hippo, and was perhaps another sign that they were becoming scarcer. It is also surprisingly difficult to ascertain the gender of infant hippos, so acquiring a female to order would not have been an easy task.

Meanwhile in London, a publicity campaign was in train to whip up expectation in anticipation of Obaysch's arrival. For example, *Punch* offered a piece called 'The Coming Animal':

---

with Obaysch – can still be seen in the Brooklyn Museum. There is a wonderful portrait of Abbott by Thomas Hicks. This shows him in full Egyptian dress puffing away at a hookah while in the background a camel train winds past the sphinx and the pyramids. Apparently he wore his Egyptian costume to one of his meetings with Whitman.

15 'London Zoological Society', *Worcestershire Chronicle*, 10 April 1850.

1 The Life and Times of Obaysch the Hippopotamus

A hippopotamus is waiting in Alexandria, to be shipped over to England. This will be the first visit ever paid to the country by this noble and rare creature. Apartments have already been engaged for him at the Zoological Gardens, where an artist will wait upon him at the very earliest opportunity with the view of taking his portrait.

A ship has been put at the disposal of the Hippopotamus; and the Captain has received orders to pay him every possible attention, and to spare no expense in 'going the whole animal.'[16]

This piece ends, presciently as it turned out, with a comment that the zoo's rhinoceros has had its nose put out of joint by the thought of this new arrival. Behind this frivolous piece is a carefully contrived strategy to manage the Obaysch project. The Zoological Society was looking to the acquisition of a hippo as a circuit breaker for its business model and no detail was ignored.

When May came round, arrangements were made to transport Obaysch to Alexandria for embarkation on a steamer to England. It was now nine months since he had been captured and he was growing fast. He would not enter the horse box which was provided for him and after consideration by Hamet a padded cart was constructed, in which Obaysch travelled the 130 miles (210 kilometres) or so to Alexandria. A huge crowd, estimated at 10,000 people, turned out to see him; troops had to be provided to control them and to escort Obaysch safely to the port. This far north, so far from the hippopotamus ranges of the White Nile, the appearance of a hippopotamus was almost as unusual as it was in London. So Obaysch had his first taste of being a 'sight' before he had even left Egypt.

On 8 or 9 May Obaysch was loaded onto the S.S. *Ripon* of the Peninsular and Oriental Steam Navigation Company under the command of Captain John Moresby. Judging by the reference to the 'liberal co-operation' of the line in the zoo's *Guide*, I suspect

---

16   *Punch*, April 1850.

that Obaysch may have been given free passage. The company's managers would have seen the huge advertising opportunity if such a rare and precious cargo could be delivered safe and sound. The *Ripon* was a relatively new ship, having been built for P&O in 1846. She was a handsome vessel: 231 feet (70 metres) long and displacing 1426 tons (1449 tonnes), her three masts stood proud above her two funnels and the immense semi-circular casings of her side wheels were painted a gleaming white, contrasting smartly with the darker livery of her hull and upper works. By 1854 she would have another important task to perform: shipping the Grenadier Guards to the Crimea. And in 1884 she brought the Italian nationalist leader Garibaldi on his much-celebrated visit to England. It was a well-run ship:

> Everything is done in regular man-of-war style. The ropes are coiled as neatly as possible, the yards are well squared and the bells sounded every half hour. The decks are well scrubbed each morning and everything is as clean and neat as can be.[17]

Into this shipshape world stepped Obaysch, to travel, as Frank Buckland put it, '*en prince*'. *Punch* (15 June 1850) said it had cost £500 to transport him, and for the next two weeks he would share the *Ripon* with 166 other passengers, 101 of whom were travelling first class. Among them was His Excellency General Jung Koorman Bahadoor Ranagee, the prime minister and commander in chief of Nepal, and his sizeable entourage, which consisted of twenty-four officers and civil servants of various ranks; Mr McLeod, the General's private secretary; and Captain Cavanagh, the British political agent in Nepal. This group was an embassy from the King of Nepal and they were carrying what was estimated as a quarter of a million pounds' worth of presents for Queen Victoria. They had

---

17   Frank Richardson Kendall, unpublished letter from S.S. *Ripon*, 19 February 1859.

1 The Life and Times of Obaysch the Hippopotamus

travelled in some style, with their expenses estimated at £20,000. This was partly because their strict Buddhism and Hinduism precluded having their food touched by anyone who didn't share their religion. To avoid this risk they booked the whole of the front of the ship and set up their own cooking apparatus, 'a large square box made of planks and paddle floats, filled with mud and sand. The fuel they used was charcoal.'

Elsewhere on the ship were a lion, a leopard, a wild pig, two pelicans, two gazelles, three eagles, three lynxes, two musk cats, an ibex, several goats and kangaroo rats, a wildcat, and a profusion of snakes and lizards. These were accompanied by 'several grim and picturesque looking Arabs and Abyssinians' including two snake charmers, who were also bound for the zoo and should be considered part of Obaysch's entourage.[18]

So, at the front of the ship was the Nepalese embassy cooking away in an oven made of wood (which sounds pretty dangerous on a ship). Somewhere in the holds was a massive collection of animals and their various exotic handlers. There were another 130 or so passengers including at least 87 others, not associated with the party from Nepal, travelling first class. And on deck, I assume between the paddle wheels at the most stable point of the vessel, stood a specially built hippopotamus house. This had been prefabricated in Southampton by the P&O line to the commission of the society. In order to accommodate a 400 gallon (1515 litre) iron tank of water, the hippo house extended into the hold below. Into this tank Obaysch could plunge by means of an access ramp. He would bathe in this tank three or four times a day for about forty-five minutes on each occasion, sometimes sinking to the bottom and other times allowing the top of his head and his back to remain above the surface. Hamet sat on a high stool in the hippopotamus house and occasionally poked Obaysch with a stick to keep him under control.

---

18 'Arrival of the Hippopotamus', *Lloyd's Illustrated Newspaper*, 2 June 1850.

Hamet continued to sleep in Obaysch's house, and one account gives a flavor of their relationship:

> Hamet had a hammock slung from the beams immediately over the place where he had been accustomed to sleep; just over, in fact, his side of the bed, his position being raised by some two or three feet. Assuring Obaysch not only by words but by extending one arm over the side so as to touch him, Hamet got into his hammock and fell asleep, when he was suddenly awakened by a jerk and a hoist, only to find himself close by the side of his compagnon du voyage. Another experiment at separate sleeping was attended by the same successful movements on the part of Obaysch, and till they arrived at Southampton, Hamet desisted from any farther trial, as he avoided in all ways any irritation of the animal. On the voyage he slept with his huge charge, who at sea especially, seemed more content, and to feel safer, when his keeper was at his side … One morning during the voyage, Hamet, from some cause or other, absented himself from Obaysch a little longer than usual, when he ran through his octave of cries, from the most plaintive to the most violent, and then was profoundly silent. 'Hamet' says the narrator, 'thought his freedom was achieved, and then, with the air of an emancipated serf, he opened his wicket and condescended to return to his tyrant – tyrant no longer, as he hoped. Hippo awaited him with a twinkle of his infant eye – that curious, prominent, versatile eye, which looks everywhere at once – as he floated in the tank, so as to command the interior of his home.' Hamet, in his great fidelity, used to keep part of his wardrobe in an angle of the roof, for convenience of making his toilet without annoying his charge by unnecessary absence. The bundle in which these choice vestments were secured had been pushed down by the vengeful infant, rubbed open with his blunt nose during that ominous silence, and, finally, left in such a state, that neither Hamet, nor any other being Mohammedan or Christian, could ever don them again.

1 The Life and Times of Obaysch the Hippopotamus

> Hamet is a well conducted Mussulman and not given to indulging in profane language, but he addressed Hippo in terms of the strongest reprobation. Hippo twinkled his eye and shook his head, blew a little trumpet through his nostrils, and smiled in triumphant malevolence.[19]

Apart from the charming, if twee, tone, this does give a credible picture of an experienced and thoughtful keeper moving a delicate baby animal. It also hints at the enormous stress on Obaysch. His range of distress calls give way to a silence during which he frantically rummages through a bundle that smells of Hamet in the hope of finding him. The tweeness suddenly loses its charm.

While all this was going on, somewhere else on the ship was a small herd of goats and cows, which had been purchased to ensure that Obaysch would have plenty to eat during his voyage. Hamet would feed Obaysch by covering his arm and hand in milk and thus he induced him to suckle. This diet appears to have worked very well as Obaysch arrived in Southampton on 25 May 1850 apparently none the worse for his trip. By June he was feeding on a porridge of milk and maize meal. We know that Hamet was already using a bag of dates to entice him to move, so we might assume (given the likelihood that he was more than a year old) that, although he still depended heavily on milk for nourishment, he was taking other foods. At any rate, it is unlikely that Hamet would have taken the risk of shipping him without ensuring that he would be able to keep him alive and the evidence suggests that *en route* Obaysch was already being fed rice.

The trip from Alexandria to Southampton took sixteen days with stops at Malta and Gibraltar, where, among other things, mail was taken on to supplement the already large consignment of letters

---

19 J. Haney, *Haney's Art of Training Animals* (New York: J. Haney & Co., 1869), pp. 149–153. Haney's account of Obaysch includes several details not found elsewhere and which appear to come from another source rather than being simply made up by Haney.

and parcels the *Ripon* was carrying from India. The 400 gallons of water in Obaysch's tank were replenished with a fresh supply daily (it was felt, probably correctly, that putting sea water in his tank might be dangerous to him) so the stops at Malta and Gibraltar must also have involved significant transfers of water. Between ports the water that condensed from the ship's steam engines was used. In sixteen days Obaysch would have needed a minimum of 6400 gallons (24,220 litres) – although this amount might be compared with the 369,000 gallons (1,396,817 litres) that today fill the 'Hippoquarium' at Toledo Zoo in the United States. Even so, Obaysch's more modest provision would still have required just under forty cubic yards of cargo space over the course of the voyage, which is a significant volume for a commercial vessel where space is money.

Three things emerge from all of this. The first is the logistical difficulty and expense of moving a large animal around the world. The second is the financial return that needed to be guaranteed in order to justify the outlay. The third is the importance of steam in the development of an international trade in wild animals. The shortened journey times afforded by the screw propeller or the paddle wheel meant that animals could be transported much more quickly and, as they were at greatest risk while in transit, much more safely and with a lower mortality rate. This significantly reduced the financial risks involved. Later the Suez Canal would shorten journey times still more, especially from Australia and India (in Obaysch's day, the quickest passenger route there from England was still a boat to Alexandria and then overland). This had an immense impact on the availability of wild animals to European collections.

When the *Ripon* docked at Southampton several things happened at once. The members of the Nepalese embassy were mortally insulted by high-handed customs officials and a major diplomatic incident started to unfold, not helped by the fact that no one had made proper arrangements for the entourage's accommodation and food. But that is another story. The everyday

1 The Life and Times of Obaysch the Hippopotamus

passengers staggered off homewards, some, no doubt, to write letters of complaint to P&O about how they were kept awake by the 115 decibel blasts of Obaysch's snoring, the roars of big cats, and the booming of ostriches. The mail was unloaded and the several weeks' old news of the births and the deaths, the promotions and the triumphs, the scandals and the disasters of loved ones in India started to find their way to every corner of the British Isles. The Noah's Ark of animals and their handlers trouped down the gangway into waiting cages on a train.

Above it all swung Obaysch.

His house and water tank, with both Obaysch and Hamet hanging on inside, were lifted by crane and placed on a railway truck, watched by a massive crowd of people who barely noticed the diamond-encrusted hats of the Nepalese princes disembarking. Then, with a blast of the whistle and a hiss and grunt of steam not unlike that of a giant hippopotamus, the special train set off for the capital, carrying Obaysch to the zoo that would be his home for the next twenty-eight years. People lined the route to see the new arrival, but he proved elusive:

> A special train conveyed him to London, every station yielding up its wondering crowd to look upon the monster as he passed – fruitlessly, for they only saw the Arab keeper, who then attended him night and day, and who, for want of air, was constrained to put his head out through the roof.[20]

He arrived late in the evening on 25 May and was tempted from his carriage and into his new pen by Hamet, who walked in front of him with a bag of dates over his shoulder. Obaysch sniffed at it as he went. The pen that had been prepared for him had its own pool and a straw-covered sleeping room, where a large stuffed sack served

---

20 *Advertiser*, 24 November 1900.

as a mattress or pillow. Hamet lived in the pen with his important charge.

Obaysch made his first public appearance the next day, which was a Sunday. This was fortunate for him, as it meant he would only be seen by the relatively small number of fellows (although this could well mean several hundred people), rather than a huge crowd, as Sunday had not yet become one of the Zoological Gardens' open access days. With Hamet at his side – Obaysch seems to have formed a very close attachment to his keeper, following him around like one of Konrad Lorenz's geese or a victim of Stockholm Syndrome – he took his first plunge into the pool. Everyone agreed that he was in good shape and 'gave satisfactory evidence of the care which had been bestowed on him, and the foresight with which the society's arrangements had been laid for its reception.'[21]

To people who had not seen a live hippopotamus before, Obaysch appeared massive. There were a few visitors, like the explorer and hunter Roualeyn George Gordon-Cumming, who had seen many, mainly by squinting at them through the sights of a heavy-calibre rifle. Gordon-Cumming's hippo trophies would shortly be shown at the Crystal Palace and, subsequently, toured around the country as the South African Museum. In fact, *Punch* commented that the height of extravagance was Gordon-Cumming paying his shilling to see the zoo's new hippo.[22]

Obaysch was by now seven feet (2.13 metres) long and six feet six inches (1.98 metres) round the middle. He had dark brown eyes and was a reddish brown in colour, periodically becoming even redder as his bloodsweat secretions came and went. The skin immediately around his eyes was much paler, more like the colour of European skin. The sound he made was described as a snort, which he would make four or five times in succession, followed

---

21 'The Hippopotamus in the Gardens of the Zoological Societies, Regent's Park', *Illustrated London News*, 1 June 1850.
22 'The Height of Extravagance', *Punch*, 31 August 1850.

1 The Life and Times of Obaysch the Hippopotamus

by a bark. This reminded people of the noise made by horses and Hamet interpreted it as a request to be taken into his pool. Most of the descriptions of Obaysch from this time concentrate on his size and his conduct in this pool – all seemed to agree that you had not really seen Obaysch until you had seen him swimming. He was compared variously to a porpoise, an otter and a seal. In the pool, the clumsy and vaguely comic lumberings he displayed on dry land were replaced by a species of elegance, demonstrating that 'the beautiful adaptation of structure to peculiar habits is in no animal more beautifully conspicuous than in the Hippopotamus.'[23]

As the society had hoped, Obaysch was an immediate sensation. People flocked to the zoo to see him. Hamet would bring him out of his pen for his several swims each day, and Obaysch must quickly have accustomed himself to the vast crowds that thronged around and jostled to get a view of him. According to 'The Diary of the Hippopotamus' published in *Punch* (6 July 1850), he rose at six in the morning, ate a bucketful of maize porridge, then promptly fell asleep again until ten o'clock, when he went out into the yard adjoining his pen to meet his public. Then he fell asleep, looking like 'an immense ball of india rubber' for most of the rest of the day. On waking in the afternoon he spent two or three hours in his pool. At six o'clock he went back into his pen, had another bucket of porridge, and went to sleep for the night. Although this description is taken from a humorous account of Obaysch's life (as we shall see, he featured regularly in *Punch* for the first year or so of his captivity) I believe that it coheres reasonably well with more official accounts of Obaysch's routine, such as that of Professor Richard Owen. Owen, described as 'the most distinguished vertebrate zoologist and palaeontologist but a most deceitful and odious man', presented an account of Obaysch's daily round in a report to the *Annals of Natural*

---

23   'The Hippopotamus in the Gardens of the Zoological Society', *Illustrated London News*, 1 June 1850. 'The Hippopotamus', *Times*, 21 May 1850, offered a kind of primer of hippo zoology and history, as if to prepare people for their encounter with Obaysch.

History. This became the most widely available serious account of the zoo's new acquisition as it was republished in a number of newspapers and periodicals. Owen made his report in the very first days of Obaysch's confinement and *Punch*'s amusing article, published about a month later, clearly draws on Owen's account for some of its details.[24]

In reality the advertisements for Obaysch (such as that in the *Sunday Times* for 2 September 1850) said that he was to be seen between eleven and four o'clock and that those who wanted to see him in his pool needed to get there early. *Knight's Cyclopedia of London* for 1851, describing the experience of seeing Obaysch, reinforced the suspicion that, curiousity though he was, he wasn't all that much fun:

> But the great attraction of 1850 has been the hippo. The town has crowded to see that rarity of Africa, which has not been exhibited in Europe for sixteen centuries. This huge prize hog with a broad muzzle disappointed public curiousity in some degree. He was asleep when some of the eager visitors wanted to see him bathe, and groping in his pond when others were anxious to see him at play with his keeper. The wonder is becoming stale. If he grows into a mighty hippopotamus such as the Romans gazed upon, he will again be popular; but perhaps he will have instinctive pining for the reeds of the Nile, and die of porridge and the washing-tub.

---

24 For example, in the *Bury and Norwich Post* (12 June 1850) and, in synopsis, in the *Leader* (8 June 1850) and the *Lady's Newspaper* (8 June 1850). Details of some of Owen's less scrupulous activities can be found in D. Cadbury, *The Dinosaur Hunters* (London: Fourth Estate, 2000) and B. Maddox, *Reading the Rocks* (London: Bloomsbury, 2017). Owen was also a convinced opponent of Darwin and the phrase 'deceitful and odious' was used of him by R.B. Freeman in his *Charles Darwin: A Companion* (Folkstone: Dawson, 1978), where he observes that 'Owen was probably the only man that Charles Darwin hated, if he could hate.'

1 The Life and Times of Obaysch the Hippopotamus

This may be brutal but it makes the point well. A shilling was a lot to pay for a daytime visit to a nocturnal animal. Such accounts contrast with the more official accounts published in the *Times* and other newspapers, where the information and often the editorial copy came direct from the Zoological Society, which was spinning Obaysch for all it was worth. In the official *Guide* even the society admitted to its visitors that a day spent at the zoo might be one 'in which, as often happened, they failed to see the hippopotamus', but recommended the rhinoceros instead. Knight also picked up on the fact that porridge was no diet for a hippo, and that a large free-ranging animal needed more than a small pool to swim in.

Obaysch got up early and then, it was reported, snorted as a request to go to his pool. Obaysch followed Hamet 'like a dog close to his heels' and then swam for half an hour or so until Hamet called him to come back into his pen, where he was fed his porridge. *Punch* referred to Obaysch's fondness for Hamet, claiming that Obaysch slept with his head in Hamet's lap and his legs thrown round Hamet's neck. Owen observed that Obaysch 'is ever impatient of any absence of its favourite attendant, rises on its hind legs, and threatens to break down the wooden fence by butting and pushing at it in a way strongly significative of its great muscular force.' Both Owen's report and *Punch's* satire agreed on two things: Obaysch was only displayed for a short period each day, and he spent a good deal of time asleep. Animals, especially nocturnal animals, do, and most visitors to a zoo will be familiar with the slight disappointment felt on discovering that most of the animals have hidden themselves away for their daytime naps and that those who haven't (meerkats, tamarinds and lemurs excepted) often look as if they might fall asleep at any moment. As *Punch* (13 July 1850) observed:

> Everybody is still running towards the Regent's Park for the purpose of passing half an hour with the Hippopotamus. The animal itself repays public curiosity with a yawn of indifference, or throws cold water on the ardour of his visitors, by suddenly

plunging into his bath, and splashing everyone within yards of him.

Nevertheless people turned out in huge numbers to see him. A letter republished with the sheet music of Louis St Mars's 'Hippopotamus Polka' (which will be considered in a little more detail in a subsequent chapter) describes a scene in which:

> a vast concourse of people of all kinds were pressing for admission – such crowding, such squeezing, I never witnessed even at the opera on a Jenny Lind night.[25]

It took nearly an hour to get to the front of the queue, whereupon the patient visitor was rewarded by the sight of an 'ugly' animal. Richard and Caroline Owen were early visitors and Caroline noted in her diary that:

> There was a dense mass of people waiting to get inside and the whole road leading to that part of the Gardens was full of a continuous stream of people.[26]

At the end of fourteen weeks the *Westmoreland Gazette* (14 September 1850) reported that 227,938 extra visitors had passed through the turnstiles, and that the zoo had made an extra £11,349 from the extra shilling they each paid to see Obaysch. In 2016 terms this represented an economic value of just under £21 million. A few

---

25 'The Hippopotamus Polka' was composed by Louis St Mars and published in London as sheet music with various additions such as this letter in 1852 by Charles Jeffreys. It can still be heard, played on piano by Gayle Glenn, at http://vimeo.com/87360673.
26 Caroline Owen, diary, quoted in Blunt, *The Ark in the Park*, p. 110. Ito, in *London Zoo and the Victorians*, p. 121, also quotes a passage from Henrietta Halliwell-Phillips's diary. She visited Obaysch on 22 June 1850 and described him as 'a short thick heavy animal something like a pig about the mouth & head & of a dark brownish colour'.

1 The Life and Times of Obaysch the Hippopotamus

months later, the *Coventry Herald and Observer* (17 February 1851) put a figure of £6186 on the extra takings just for that year to date. The economic significance of Obaysch to the zoo is mentioned again here solely to show how Obaysch's life was now defined by a massive gaping crowd.[27] His relatively solitary existence as a free ranging, free swimming hippopotamus was now replaced by a small pool and a pail of porridge, and by ring after ring of craning faces and wide open human mouths.

In the first fourteen weeks he would have been on display for 490 hours, which means that an average of 465 people per hour (not counting celebrity visitors such as Queen Victoria and her family, who would have been able to visit out of hours) crowded round his pen. The Queen first wrote about Obaysch in early June in a way that shows she was already familiar with the press reports and was deeply interested in him:

> It is said to be the first [hippo] that has come to Europe since Roman times. It is a very sagacious animal and so attached to its keepers that it screams if they leave it and the man is obliged to sleep next to it.[28]

Her first visit was on 18 July 1850, when she was accompanied by her daughters. They made a beeline for Obaysch:

---

27 See also Ito, *London Zoo and the Victorians*, pp. 80–137.
28 Queen Victoria quoted in the *Courier-Mail*, 20 March 1952, in an article about the Landseer watercolours of Obaysch that she had bought for the royal collection. The second passage is also quoted by Blunt, *The Ark in the Park*, pp. 114–115, who reproduces an image of the two snake charmers taken from an 1850 issue of the *Ladies' Companion*. Victoria bought a set of six photographs of animals at the zoo, including the one of Obaysch shown on the front cover. Alas, in Dublin, her statue was known as 'The Hippopotamus'. See M. Homans and A. Munich, *Remaking Queen Victoria* (Cambridge: Cambridge University Press, 1997), p. 236.

We were received by Mr Mitchell, the Secretary, and went straight to the house where the hippopotamus is kept. We had an excellent sight of this truly extraordinary animal. It is only 10 months old and its teeth are just coming through. Its eyes are very intelligent. It was in the water, rolling about like a porpoise, occasionally disappearing entirely. The little Egyptian serpent charmer enticed it out by grunting to it, and holding out a piece of bread.

The royal party then went on to see the snake charming display outside the giraffe house (this is dealt with in more detail below) and from her description we learn that one of the snake charmers, Merwan – who was later to return to the zoo with a different job – was only fifteen years old. The note above suggests that Merwan was also helping Hamet to care for Obaysch. The Queen enjoyed her visit so much that she made five more in quick succession.

Obaysch must have been terrified. Only the comforting and familiar presence of the kindly Hamet offered a point of contact with anything he had known before.

The regular contributor to *Fraser's Magazine* who wrote under the name of 'The Naturalist' recorded a visit in 1850 that is worth citing in full, as it offers a number of details not found elsewhere:

> Hippo was in his bath. When he sinks he puts back his ears and closes them to keep out the water. A large vegetable marrow was thrown to him by Hamet. He mumbled it for some time in the water, and below the surface as well as above, making an impression on the fruit but not breaking it. When below the surface he would let it out of his mouth, and then rise after it as it floated to the top, trying his young teeth upon it. At last his vegetable appetite appeared to be roused. He brought it to one of the steps of his bath, and, reposing, set to work on it in good earnest, with all but his head still in the water, succeeded in breaking it, bit off pieces, chewed them with a slow, champing, snapping motion, without any lateral grinding, and swallowed

them. He had previously been offered green maize, which he mumbled, broke, and played with, but did not eat, so far as I could see. Boiled carrots and kohl-rübe [sic] were more to his taste; and he had eaten freely of them before the experiment with the raw vegetable marrow was made. All this looks like a healthy state of stomach, and I cannot help hoping that his careful attendants will bring him through the winter. He was rather fractious at first when left, but is now reconciled to the absence of his kind Hamet at night, and sleeps by himself very comfortably. In short, his conduct entirely justifies the epithet conferred on him by Mr Dickens, who has immortalized 'The Good Hippopotamus'.[29]

It is interesting to note that 'The Naturalist' was concerned even at this early stage that Obaysch might not survive a northern European winter. In 1851 he visited again and we learn a little more:

Hippo very much grown and thriving admirably. His food still oatmeal and milk, and it must be told – as the well-bred Hamet informed me in a whisper 'many horse-dung', of which latter condiment he consumes a great deal and has long done so.[30]

'The Naturalist' then speculates, accurately, that this dung regime may be helping Obaysch to build his capacity to digest milk. Hamet clearly knew what he was doing. As we shall see later, undigested milk was frequently found in the guts of young hippos born in other European zoos later in the century after they died in infancy. The fact that the professional naturalists appear not to have picked this up suggests that they did not ask Hamet's opinion as much as they might have done, or, if they did, that they did not value it. One

---

29 'Leaves from the Notebook of a Naturalist', *Fraser's Magazine* 42 (July–December 1850), pp. 196–202.
30 'Leaves from the Notebooks of a Naturalist', *Fraser's magazine* 43 (January–July 1851), pp. 546–547. In a later place The Naturalist mentions a new diet for the growing Obaysch: clover chaff tea.

anecdote describes how Hamet expressed disbelief of Herodotus and Aristotle when he was told what they had described a cleaning relationship between crocodiles and certain Nilotic birds. But he then said that he knew which birds they meant, and was readily able to point out one of the birds in question, the spur-winged lapwing *Hoplopterus spinosus*, when David Mitchell took him to the aviary. In fact, Hamet was correct: the cleaning behaviour Herodotus attributes to a bird he called 'trochilus' (which has since been identified as *Hoplopterus spinosus*) has never been observed.

An early visitor from France saw other qualities in Obaysch, which he recorded in a piece in *La Sylphide* in December 1851:

> *Pendant notre séjour à Londres, nous avons rendus de fréquentes visites à Obaysch, étudé avec intérèt ses habitudes; nous avons remarqué son intelligence qui est très grande et surtout l'amour qu'il porte à son gardien Hamet.*

> [During our stay in London we made frequent visits to Obaysch, studied his habits with interest; we noted his intelligence, which is very great, and, above all the love which he bears towards his keeper Hamet.]

It is rare to find commentary that sees Obaysch as an intelligent animal. But the combination of praise for Obaysch's rational faculty and his affection for Hamet perhaps locates this view of him within a specifically French way of seeing animals, which was somewhat different from the Anglo-Saxon approach.[31] One sees this clearly

---

31 There is a study yet to be done on the comparison between French and English attitudes to exotic animals in the nineteenth century. A good start can be made from the following: C. Crossley, *Consumable Metaphors: Attitudes to Animals and Vegetarianism in Nineteenth-Century France* (Oxford: Peter Lang, 2003); K. Kete, *The Beast in the Boudoir: Pet-Keeping in Nineteenth-Century Paris* (Berkeley: University of California Press, 1994); L.E. Robbins, *Elephant Slaves and Pampered Pets: Exotic Animals in Eighteenth-Century Paris* (Baltimore: Johns Hopkins University Press, 2004);

1 The Life and Times of Obaysch the Hippopotamus

in the depiction of Pacific island and Australasian animals during the late eighteenth and early nineteenth centuries. The French illustrators are far more concerned than the British with the context of the animals and, in particular, offer images that bring out an essentially Romantic view of the animals' inner lives – for example, compositions that stress familial bonding and affection.[32] In addition, there was clearly a French tradition – which one sees in the case of the lion who was liberated by revolutionaries from the royal menagerie and became one of the first inhabitants of the Jardin des Plantes – of attributing to animals a greater ratiocinative power and inner life than tended to be the case where naturalists working in the British context were concerned.[33] It may come down to a traditional faultline between theoretical and empirical approaches that can still be traced in some of the differences between French and English academic discourse today. As late as July 1874 the French magazine *Journal des Haras* published a long memoir of a visit to Obaysch when he was a newcomer to the zoo, which went into great detail about his close and affectionate relationship with Hamet.

The American naturalist Zadock Thompson recorded a visit to see Obaysch and, from him, we get a slightly less genial view of the display:

> I found it [Obaysch] confined in a yard – perhaps four rods square [about twenty-two by twenty-two yards, or eighteen by eighteen

---

I. Tague, *Animal Companions : Pets and Social Change in Eighteenth-Century Britain* (Philadelphia: Pennsylvania University Press, 2015); and the essays in the first section of *French Thinking about Animals*, edited by L. Mackenzie and S. Posthumus (East Lansing: Michigan State University Press, 2015).

32 See J. Simons, *Kangaroo* (London: Reaktion, 2013), pp. 91–96 for some commentary on this. The most comprehensive recent study of animals in British pictorial culture (1750–1850) is D. Donald, *Picturing Animals in Britain, 1750–1850* (New Haven, CT: Yale University Press, 2007).

33 J. Simons, 'Radical Animals: Exotic Animals in the French Revolution', unpublished paper delivered at 'Liberty and Limits' Conference, Macquarie University, 5–6 December 2013.

metres] – with a pond or basin of water in the centre about two rods across, and the inclosure was surrounded by two or three hundred spectators. The hippopotamus was lying on the platform by the side of the pond, with its eyes closed and apparently asleep. The Nubian keeper soon roused him from his slumbers, and drove him into the water. He waded as long as he could touch bottom, and then swam lazily across, and crawled partly out on the other side; but he was driven back again into the water, and, after remaining there some little time, nearly motionless, he was permitted to come out from the side he went in.[34]

As we saw in Knight's commentary, accounts of Obaysch that were not controlled by the Zoological Society or its supporters appear to describe more coercion than the image of an imprinted little calf trotting like a puppy after its keeper would suggest. Obaysch was a nocturnal animal used to sleeping almost all day. He is doing just that when he is woken up and forced to go into the water. He tries to get out but is 'driven' back. So he goes to sleep in the water. The frequent claim that to appreciate Obaysch you had to see him swimming (in fact, hippos can't swim: they punt themselves along the river bed with their slightly webbed toes) means Obaysch is forced to swim when he doesn't want to. In order to justify the extra shilling that hundreds of people were paying each day to see him, the Zoological Gardens had to stage a performance, in which Obaysch was required to do everything that went against his experience as a wild animal in Africa. Another early account in *Sharpe's London Magazine* (1850) perhaps hints at underlying problems that would later come to the fore:

---

34 Z. Thompson, *Journal of a Trip to London, Paris and the Great Exhibition* (Burlington: Nichols & Warren, 1852), p. 90. Other American sightings of Obaysch at this time can be found in W.A. Drew, *Glimpses and Gatherings* (Augusta, GA: Homan & Markey, 1852) and D.W. Bartlett, *What I Saw in London* (New York: C.M. Saxton, Barker & Co., 1852).

1 The Life and Times of Obaysch the Hippopotamus

in its play [the hippopotamus] (think of a seven foot long sow being fond of a joke) jumps at its keeper with its comical as yet toothless mouth and alligator like muzzle, but takes care not to touch him.

This is the only journalistic account that picks up the detail that Obaysch has yet to grow his teeth, but the hint is there that when they come in the play will be less funny. I also wonder if Obaysch took care not to touch Hamet because he had learned that doing so earned him a heavy poke from that stick.

The Zoological Society was very keen at this time to differentiate its gardens from the more popular zoos and menageries that were springing up all over Britain as the spread of empire facilitated the acquisition of exotic animals. It did this by stressing the zoo's scientific and educational aspects. It was important for the society to maintain, wherever possible, that what you saw in the Zoological Gardens were animals behaving as they would in the wild or, to put it another way, as God had created them. When one looks at the exhibition of an animal such as Obaysch, it becomes clear just how much artifice was involved in this claim. Although it wasn't a crude as the elephant or camel giving rides, the llama pulling a cart full of children or the chimpanzees' tea party, it was nonetheless a performance. It required a tired nocturnal animal to drag himself up and down a small pond when all he wanted to do was have a sleep. *Punch* (10 August 1850) summed it up with a cartoon captioned 'Brown, Jones and Robinson go to the Zoological Gardens'. The three friends do various exciting things and then they come to Obaysch. Three heads peer over the fence at a small hippo, who is lying on the ground pretending to sleep while peering warily and unhappily at them from one eye.

In 1851 the Kilvert family went to see Obaysch, and we get from Emily Kilvert a rare account of what it was like to be in the crowd:

> When [the hippo] came out of its tank, Frank [her brother, who would grow up to be the famous naturalist and diarist the Reverend Francis Kilvert] naively asked where his bath towel was, at which the people standing nearby tittered a good deal.[35]

Even in 1855 people still flocked to see Obaysch in the water:

> The hippo surges into his bath in the inclosure ... and there is a rush of visitors to see the mighty brute perform his ablutions.[36]

The word 'surges' captures precisely that moment when Obaysch, who has been penned up all night away from his pool, comes charging through the gates of his pen and into the water while, in the other direction, a crowd comes running to see him. This is perhaps the only description that authentically captures the dynamics of a visit to see Obaysch and the relationship of constant movement between the hippo and his audience. When he was asleep the crowd stood still. When he moved the crowd moved.

For the first year of his life Obaysch was the undoubted star of the zoo and the biggest draw card for visitors to Regent's Park. This changed within a year: in 1851 a baby elephant was born, which, as the *Illustrated London News* put it, promised 'to be nearly as popular as the Hippopotamus acquisition of last year'. A giant tortoise, a gift from Queen Victoria that Richard Owen went to pick up from Buckingham Palace, where he rode on its back, was installed as the next star animal. It didn't survive the winter but was the subject of a *Punch* cartoon in which it and Obaysch were depicted as 'those distinguished "lions", arriving in evening dress at

---

35 J. Toman, Kilvert's *World of Wonders: Growing Up in Mid-Victorian England* (London: Lutterworth Press, 2014), p. 78. Later in his life Kilvert appears to have been a champion of nude bathing. Perhaps the experience of seeing a hippo without a bath towel marked him for life.
36 A. Wynter, *Curiousities of Civilisation* (London: R. Hendricks, 1860), p.119. Wynter was writing in 1855.

a society ball (1 January 1851). In fact, the baby elephant attracted much more attention than the tortoise. *Punch* (10 May 1851) picked up this reaction and published a cartoon captioned 'The nose of the hippopotamus put out of joint by the young elephant': while Obaysch sobbed, Hamet comforted him: 'Nebber mind den! him sall be a lubbly 'potamus for all um great ugly elflint.'

On 31 May *Punch* published 'The Lament of the Hippopotamus', a poem in which Obaysch mourns the loss of his popularity but hopes for better things to come:

> But let the clumsy infant its triumph now enjoy;
> The brute must quit its babyhood, and cease to be a toy,
> Oh, then, farewell forever, its glories and its charms!
> It can't remain forever an Elephant in arms.
> 'Tis then the Hippopotamus will reassume its sway,
> Growing in popularity, as well as size, each day.
> Glories will light upon our race – the public will allot 'em us;
> Then, hip! hip! hip! hip! hip! hurrah! hip! hip! for the Hippopotamus!

As early as 5 October 1850, *Jackson's Oxford Journal* had published some verse entitled 'The Lamentation of the Hippopotamus' in which Obaysch reflects on the downturn in his popularity now that the London season is over.

But, in June 1851, the planned extension to Obaysch's accommodation was made and he acquired a new pool. It was deep enough to accommodate the adult animal he was all too quickly becoming and it was thirty-three feet (ten metres) square. This would not have compensated for the loss of 275 yards of free swimming space in the Nile, but was a big improvement on the small semi-enclosed pool he had hitherto occupied. The extension was the first phase of a program of improvement and enlargement to the hippo accommodation that would go on periodically throughout Obaysch's life. The *Report of the Council of the Zoological Society* records modifications in 1854, 1862, 1863, 1867, 1868, 1869, 1871 and 1872 (when, among other things, stronger iron bars and an iron front were fitted). *The Illustrated London News* (14 June 1851) reported the opening of the bath with somewhat different priorities in mind:

> All the inconveniences to which visitors were subjected last year from the bath being within doors are now obviated, and the platform affords good accommodation for about a thousand spectators at the same time.

An engraving that accompanied the article shows Obaysch largely submerged and forging though the water while Hamet, wearing Egyptian dress including a tarboosh and baggy trousers and holding a pointed stick, watches on solicitously. A giraffe peers curiously over the bars of an adjoining cage. Obaysch's pool is surrounded by iron railings and trees that he would not have recognised. In every spare inch of space outside, fading into the distance, are the round top hatted or bonneted faces of the Victorian public drinking in the marvels of their shilling's worth. This new platform was replaced in 1862 when the Council noted in its *Report* that 'the wooden stages at the south and east sides of the hippopotamus pond have become rotten and unsafe' and replaced them with brick. New seats were installed in 1863. These repairs and improvements demonstrate that over a decade after his arrival in London Obaysch continued to attract visitors in sufficient numbers to justify an ongoing investment.

An American traveller saw Obaysch in 1852 (i.e. after he had moved to his new accommodation) and described the exhibition as follows:

> He resided in a separate establishment, having a large bath fitted up for his especial accommodation. A raised platform, about three or four feet wide, extended the length of the room; and on this we located ourself, in company with three or four visitors, and the Nubian keeper. At first nothing was to be seen but a slight

rustling in the water, and then a huge proboscis was thrust up as if to sniff the summer air, and then hidden again. By and bye there was a great plunging, and sleek and shiny our young friend appeared upon the surface, swimming like a gallant cruiser in his own confined ocean. Then he went down again, and then made up his mind to land, for putting one great paw upon the stone steps that ascended from his bath, out he came with a snort, all dripping wet, and went about his paddock as if greatly refreshed.[37]

There are interesting differences between this account and those cited above. Firstly, although the performative element remains, the coercive element is much reduced. Obaysch swims when he likes and stays in or comes out of his pond when he likes. Hamet stays safely on the platform with the spectators (and the engraving which accompanies the article shows him watching Obaysch emerge from his pond from behind the iron railings). It is also worth noting that this exhibition was witnessed by only a handful of spectators, so one wonders if some arrangement for a private viewing had been made. Many zoos today offer premium access to their star animals. For example, Adelaide Zoo currently offers a 'VIP Panda Experience', where for $500 you can help Wang Wang and Fu Ni's keepers to feed them. It seems likely that the Zoological Society would have made efforts to maximise their returns in this way, although I have found no evidence of this apart from the observation, quoted above, that one day in 1852 there were only four visitors to an otherwise overwhelmingly popular attraction. Obaysch is now depicted as significantly bigger than in earlier images. It is clear what has happened: Obaysch is growing up and you cannot, whatever your animal whispering skills, get away with poking a large male hippo with a stick. Obaysch's innate size and fierceness were starting to enable him, to a small extent, to live a life according to his own rhythms, which very dimly shadowed those he had been taught by his mother nearly four years before and far away down the White Nile.

---

37 Anon., 'The Hippopotamus', *Illustrated Magazine of Art* 1 (1853), pp. 80–82.

Obaysch had now been in the zoo for over a year, yet his popularity was such that it was still worth building a viewing space capable of holding a thousand people at a time. That would bring in one thousand shillings (or fifty pounds) whenever it was full. Let's say that it was open four hours per day and the average person spent half an hour watching Obaysch. That would bring in £400 per day, which was a very considerable sum at the time and shows how dependent the zoo's finances had become on Obaysch. The new addition to Obaysch's pen and the better opportunities to see him it afforded were also important because 1851 marked the opening of the Great Exhibition and the 'all-absorbing Crystal Palace'. To compete with this national attraction the zoo had to put on a good show and this it certainly did: a new pool for Obaysch, a baby elephant, and a wonderful display of over 320 different species of hummingbirds, stuffed, mounted in twenty-four cases, and, initially, financed by John Gould. This attracted another 75,000 visitors, each of whom paid an extra shilling to see it. Charles Dickens described the show a magnificent spectacle, which should encourage people to 'study ... the work of the Divine hand in these little birds.'[38] Queen Victoria and Prince Albert were very much enamoured of this marvelous display, and what with their patronage and the warm approval of Dickens it was little wonder that Gould and the Zoological Society netted £3750 from the enterprise.[39]

In order to understand the popularity of Obaysch and the hummingbirds a comparison might be made with 'Giraffe Heights', a viewing platform built at the Zoological Society's Whipsnade Zoo in 2013. It can accommodate 300 people at a time and the tone of the society's publicity clearly suggests that this is a very good number indeed. Of course, you don't have to pay extra to access the platform, as you did to see Obaysch, and the real economic value of an entry

---

38 *Times*, 17 May 1851. Six of Gould's hummingbird cases are still in the Natural History Museum and one may be in a private collection.
39 Ito, *London Zoo and the Victorians*, pp. 132–133.

1 The Life and Times of Obaysch the Hippopotamus

ticket to Whipsnade is a fraction of what it cost to get into the zoo and to see Obaysch in 1851.

The Zoological Gardens clearly had little to fear from the Great Exhibition. People came from all over the country to see the Crystal Palace and its contents (including a virtuoso display of false teeth made from hippopotamus ivory by Mr Finzi). All manner of special deals and packages were available to all but the very poorest, and having made the journey and paid for the accommodation, the visitors would naturally want to see other things. And so they made their way to the zoo, some of them guided by the special souvenir gloves that were being sold with street maps of London printed on the palms. Enhancing the experience of zoo visitors might be seen not so much as a ploy to compete with the Crystal Palace as a way of managing the continuing, and even increasing, high numbers that the vast influx of people to London would bring to the zoo. Gould's hummingbird exhibition was a clear sign that the Zoological Society eventually came to believe that the Great Exhibition represented an opportunity rather than a threat. And so it proved to be.

Obaysch was now over two years old and should have been able to enjoy at least one more year of being looked after by his mother. Instead, he spent his days surrounded by thousands of spectators, with only Hamet and the various dignitaries of the Zoological Society and invited naturalists and scientists offering a more intimate level of contact. 'Hippopotamus Murray' used to drop in, and would speak to Obaysch in Arabic, a habit he apparently continued throughout Obaysch's life. He did after all owe Obaysch some gratitude for his own fifteen minutes of fame, as he recorded in a diary he kept while he was British ambassador to the court of the Shah of Persia:

> So there are lions here, despite my incredulity. So there are also in London, Paris, and everywhere else! I was one myself for a short time – i.e. when I brought over to England the first hippopotamus

that had been seen alive in Europe since the earlier days of the Roman Empire, and I was called Hippopotamus Murray.[40]

Obaysch was by now eating about 'a hundred pounds of hay, chaff, corn, roots and green food' every day. He was surrounded by the noises and smells of other animals. Some, such as crocodiles, he would have recognised; others, such as wombats, would have puzzled him. He would have seen very few of them – although the giraffe house, an impressive and rather beautiful building designed by Decimus Burton, was next door to his pen and he could have seen their heads raised above the bars of their enclosure as they stalked gloomily up and down. He was described in 1852 as:

> lazily reposing on the side of a pond, which, he finds it is requisite every now and again to lave his huge and somewhat ungainly carcass ... The Hippopotamus is at present the pet of the Gardens, and (what is worse) he seems to know it. Nothing can exceed his heavy, contented, stupid looking face. His eye is prominent, his cheeks are blown, his movements lazy and he is prone to indulge in a siesta.[41]

The pejorative use of the word 'siesta' is interesting here. Most obviously it implies that Obaysch is, like all foreigners, lazy. What it really records, however, is his natural tendency, as a nocturnal animal, to sleep during the day.

---

40 Murray's Persian diary quoted in M. Gail, *Persia and the Victorians* (London: Routledge, 2013), p. 45. Alas, the diplomatic skills that Murray had displayed in the acquisition of Obaysch were not sustained during this posting and it is arguable that his failure to understand the Shah led to the Anglo–Persian war of 1856–57.
41 *The Zoological Gardens: A Description of the Gardens and Menagerie* (London: Zoological Society, 1852), pp. 30–31.

1 The Life and Times of Obaysch the Hippopotamus

In his early years Obaysch was also part of a more elaborate Orientalist display. On the S.S. *Ripon* arrived various snakes and various exotic handlers.[42] One of these was Jabar Abou Haijab. Jabar's authority derived from his claim to have charmed snakes for Napoleon during the French invasion of Egypt over fifty years before and to have been a collector for Geoffroy Saint-Hilaire, one of the most respected naturalists of the period. Another was the much younger Mohammed Abou Merwan, whom we have already met in Queen Victoria's account of her visit to Obaysch. These two and possibly others came to the zoo with their collection of venomous serpents and settled down as Hamet's assistants, as well as offering a snake-charming show at four o'clock each afternoon.

This display achieved four things.

Firstly, it reinforced the correctness of the decision to build a snake house at the zoo. This had only opened in 1849, so the arrival of the snake charmers in 1850 drew attention to the attractiveness of the new facility.

Secondly, the snake charming show gave visitors who had come hoping to see Obaysch but who may only have caught a glimpse of him something interesting and lively to see. The show began with a procession from the giraffe house, adjacent to Obaysch's pen, before proceeding to the reptile house. The initial audience would thus have been anyone who was in Obaysch's vicinity when the procession set off.

Thirdly, the exotic show – Merwan in particular made a great deal of his spectacular Arab or Turkish costume – located Obaysch in an Oriental world full of dangers but also full of strange pleasures.

Finally, the show started at four o'clock in the afternoon, the time that Obaysch was taken off display, so one assumes that it also

---

42 See J. Hall, 'The Snake-Charmers at the Zoo', and J. Hall, 'Encountering Snakes in Early Victorian London: The First Reptile House at the Zoological Gardens', *History of Science* 53 (2015), pp. 338–361.

helped to draw the crowds away from his pen to make it easier to calm him in preparation for his evening.

There was typically a great crush to see the procession – which shows how many people were still near the hippopotamus house when it was closed for the night. No one wanted to get too near the snakes but, as William Broderip noted in an account of the show in *Fraser's Magazine* (1850), the experience of being in the crowd was a bit like being in Lars Porsena's army. The press of people at the back who were anxious to see the show pushed those in front rather closer to the snakes than they really wanted to be. They need not have worried: as Jabar admitted, the act would not have been possible without removing the snakes' fangs. What we learn from this interview was that Hamet spoke sufficient English to translate intelligibly for Jabar, who only spoke Arabic. This would certainly help to account for Hamet's success in joining Murray's household back in Cairo and his selection as Obaysch's companion and keeper. The snake charmers were not a permanent fixture – they returned to Egypt at the end of 1850, bearing a substantial gratuity of twenty-one pounds – but their presence shows how carefully the Zoological Society was managing its new star attraction, and how it wished to locate him in their public's imagination.

Most significantly, the show contextualised Obaysch as a specifically Egyptian animal and not, more generally and perhaps more accurately, as African. This makes it easier to place him within the growing discourse of Orientalism, which, before the Indian Mutiny broke the charm in 1857, held the Victorian public's gaze as surely as young Merwan or old Jabar could stare down and transfix a King Cobra.

Obaysch was now settling down into his life both as an attraction and as an animal whose celebrity was gradually fading. As 1851 came to a close and the unprecedented numbers drawn to the zoo not only by the star animals but also by the Great Exhibition started to diminish, he would now become a special but increasingly routine port of call on a walk around the zoo, rather than the must-see attraction he had

1 The Life and Times of Obaysch the Hippopotamus

been in 1850. By 1851 *Punch* and a number of other periodicals had published skits and poems in which Obaysch lamented his declining fame. *Punch* (10 August 1850) even published a piece called 'Fleeting Popularity', which reflected on a

> Gentleman called Hamet who is enjoying just now a large amount of popularity as the bed-fellow of the Hippopotamus. Unfortunately, the career of this individual hangs on a slender thread – the thread in question being the life of the animal from whom he derives all the éclat that at present belongs to him. Should anything happen to the Hippopotamus, it is too clear that Hamet will no longer be of interest.

Nasty but true. But Hamet proved a good hippopotamus keeper, who was surely responsible for Obaysch's survival. It appears that by 1853 Hamet had left the zoo, and by 1870 he was being remembered as 'Hippopotamus Johnny':

> the stout, short, jolly person who waited on the hippopotami long ago in the Zoological Gardens and who, from association with them, had become almost hippopotamic in expression – dark-skinned, dark-eyed, sleek and round. This was Hippopotamus Johnny. He came to England with the first little river-horse – 'all de same as my little child' – and left his old mother somewhere up in Nubia, in the parts near Khartoum; and he has since been dragoman for 'de long desert journey.' A devout man is Hamed, and very honest I think, and very slow.[43]

Attempts to represent Hamet's speech consistently figure him as a cartoon African, not an Arab, so we should probably assume that he was a Nubian or Sudanese who looked like an African in spite of the Oriental costume he wore when on duty with Obaysch and their shared public. Interestingly, in 1851 two of Obaysch's American

---

43 'Hippopotamus Johnny', *Kind Words*, 12 May, 1870.

visitors saw Hamet quite differently. As we have seen, Zadock Thompson described him, probably accurately, as a Nubian, while the Reverend Mark Trafton wrote, in his 1852 *Rambles in Europe*, of 'his East India keeper'. The piece cited above speaks of Hamet as being associated with 'hippopotami' but the evidence suggests that he had only ever been Obaysch's keeper.

In 1853 Obaysch makes a brief appearance in the press when his disgruntled feelings about the arrival of a new star animal, this time an 'American ant eater', were imagined in a verse published in *Punch* on 22 October 1853 that concludes:

> Or I wish that I could make myself a Fellow, d'ye see,
> Of this Zoological So-ci-e-ty:
> For then I'd send this Ant-eater back to his Ants
> Or to my French rivals at the Jardin des Plantes.
> But it's oh! no go: there's no end to my woes:
> The American Ant-eater out of joint's put my nose!

If nothing else this shows that even when there appeared to be a falling off of interest in Obaysch, he was still sufficiently famous to merit this kind of attention and this kind of emotional projection.

The drowsy routine of a slowly fading star was shattered and re-energised on 21 July 1854, when the S.S. *Ripon* drew alongside at Southampton once again – once again carrying a hippopotamus. This time it was a female. She was another gift from Pasha Abbas in fulfillment of his promise to Murray that he would provide a mate for Obaysch. Like Obaysch, she had been held for a time in Egypt; 'Hippopotamus' Murray had by then moved on and it was the new consul-general, Frederick William Bruce (later Wright-Bruce), who took delivery of the new animal in Cairo. Unlike Murray, Bruce did not have to keep the hippopotamus alive and well himself. This time the Pasha housed her in winter quarters in his own palace. The new

1 The Life and Times of Obaysch the Hippopotamus

hippo had been captured probably in August or September 1853 and arrived in Cairo in November, where she stayed pending her transportation to England in July. The details of her capture are not to be found anywhere and there is no record of her having a wound like Obaysch's, which she would surely have had if she had been hunted by the traditional method. This seems to me possible evidence that she was considerably younger than Obaysch when taken and that it had been possible simply to abduct her with or without the killing of her mother – although her size on arrival in London calls this into doubt. In any case, the *Report of the Council* for 1855 noted that:

> The experience acquired in 1850 rendered this [i.e. the capture and international transportation of a hippo] a comparatively easy undertaking …

The new hippo did not travel alone: as Obaysch had Hamet, so his putative mate travelled with Mohammed Abou Merwan, the younger of the two snake charmers who had originally come over with Obaysch. Merwan had a lucky escape, it appears, when the hippo pinned him against the wall of her pen and had to be pushed off by crew members, but he was also able to beguile her with 'a doleful and monotonous chant', as a result of which her 'great bulk vibrated to and fro, as if keeping time to the measure of the keeper's song.' She also enjoyed the performances of the *Ripon*'s band and would raise her head 'in the attitude of listening' when she heard the music. Like Hamet before him, Merwan spent his days and nights in the pen.

The *Ripon* arrived in Southampton on a Friday night, so it was not possible to unload the new creature until the Saturday. Merwan spent the time profitably. He lit up the pen with a candle and performed his hippo-charming act for the people who had come aboard to see the new arrival, before reclining on his seat to be watched by them.[44]

---

44 'Arrival of Another Hippopotamus', *Illustrated London News*, 29 July 1854.

The new hippo was almost immediately called Adhela, and she took up residence in Obaysch's old pen. She must have been not much over eighteenth months old when she arrived in London, as she is described as 'only a suckling calf', although weighing over a ton she was heavier than Obaysch had been when he reached Britain. Obaysch was now at least five years old, so he was confronted with an immature female and the first illustration of their meeting, in the *Illustrated London News* on 12 August 1854, shows a very large Obaysch looking placidly at a rather small Adhela. She is gazing devotedly at a man wearing Oriental clothing and carrying a switch. Significantly, there are bars between them and an opened gate, which one assumes could be closed to prevent Obaysch and Adhela from seeing each other. Obaysch is presumably in the much expanded pen that had been built for him to keep pace with his growth – he now had teeth and was eating corn. He had been the only hippopotamus in the zoo for four years. Indeed, he was the only hippo in Europe until 1853, when the Jardin des Plantes in Paris finally acquired a male. This was also a gift from Abbas, who was plainly balancing up his options by trying to keep the French sweet as well as the British. So when Adhela arrived did Obaysch know that he was a hippopotamus and did he recognise Adhela as being of the same species? If Obaysch knew she was a hippo like him he may not have known she was female. She was far from sexual maturity, and immature hippos are very hard to sex (by humans, at least; hippos themselves may find it easier). It must have been confusing for him – he had not seen another hippo since 1849 and now another hippo who was little more than a baby appeared from nowhere while he was growing into massive adolescence. He had about two years to go before he was sexually mature. Adelha had three and a half to four. So before him stretched an awkward two years. The society did not, however, realise this. Knowledge of the breeding cycles of hippos was still so minimal that they believed that 'there is now some probability that in the course of next year

1 The Life and Times of Obaysch the Hippopotamus

the Hippopotamus will be added to the list of species which have reproduced in the establishment.'[45]

So it was to cope not only with the increasing size of Obaysch but also with the need to maintain two hippos at different phases of maturity that the Zoological Society embarked on the building of a bigger hippo house. This contained a bath that was thirty-five by fifteen feet (10.6 by 4.6 metres) and nine feet (2.74 metres) deep. The whole thing was fenced in with what were described as 'massive iron railings commensurate with the enormous force which the animal is rapidly attaining.' There must have been some show of this force to justify yet another build beyond simply keeping pace with Obaysch's growth. Adhela was always going to be much smaller than Obaysch – although this probably wasn't realised at the time – but there is evidence that she was a very aggressive animal. She once attacked someone standing behind the iron bars of her pen and bit so hard that she snapped off one of her teeth. In spite of the hint quoted above that Obaysch was also forceful – of course he was – there are no explicit reports of that kind of aggression. However, after Bartlett's death his son published a volume of memoirs in which his father spoke of Obaysch's gentle nature when he arrived in the zoo. But later he encountered Obaysch in his fiercer mode:

> Well, for some reason or other the brute got attached to me, I believe it was because I talked to him whenever I saw him. We were the greatest of friends and he was so docile I used to ride on his back. In 1852 [when Obaysch was only three or four years old, so very far from fully grown] I was engaged in mounting a specimen hippo for the Crystal Palace and went into Obaysch's den to take some measurements. Thinking no evil, I was busy with my tape when it suddenly slipped, and the brute turned round on

---

45   *Report of the Council of the Zoological Society*, 1855, p. 22. In fact, modifications to take into account the birth of hippo babies were not made until 1872 and 1873: *Report of the Council of the Zoological Society*, 1872 and 1873, pp. 20, 22.

me with a furious snort, gnashing his jaws fiercely. I rushed for my life, and escaped through the rails, the keeper, who was with me, doing the same. It was a very near thing indeed for both of us.[46]

He added that 'from this time no one except his keeper, Hunt, would venture into his den.' He also notes that, like Adhela, Obaysch managed to snap off one of his teeth while trying to bite at someone. Adhela was always, it appeared, 'emphatically a good hater' but clearly Obaysch, like all hippos, was an aggressive and dangerously unpredictable creature.

The doctor Andrew Wynter offered one of the very few images of the Obaysch the Zoological Society didn't want people to see:

> At times, indeed, he is perfectly furious, and his vast strength has necessitated the construction of his house on a much stronger plan. Those only who have seen him rush with extended jaws at the massive oaken door of his apartment, returning again and again to the charge and making the solid beams quiver as though they were only of inch-deal can understand the dangerous fits which now and then are exhibited by the creature who was so gentle when he made his début that he could not go to sleep without having his Arab keeper's feet to lay his neck upon ... The enormously strong railings in front of his apartments are essential to guard against the rushes he sometimes makes at people he does not like.[47]

Wynter claims that he even saw Obaysch leap eight feet out of his bath to charge a workman who was strolling along the platform of his enclosure and 'the infuriated animal ... would speedily have pulled the whole construction down, had not the man run rapidly

---

46 'Hippos' Habits', *Pall Mall Gazette*, 8 January 1887. This account was replicated in Edward Bartlett's collection of his father's memoirs and papers, *Wild Animals in Captivity* (London: Chapman & Hall, 1899).
47 Wynter, *Curiosities of Civilisation*, p. 119.

1 The Life and Times of Obaysch the Hippopotamus

out of his sight.' In the same article he also mentions another incident that had otherwise been suppressed:

> This affection for his nurse has undergone a great change, for, on Hamet's countryman and coadjutor, Mohammed, making his second appearance with the young female hippopotamus, Obaysch very nearly killed him in the violence of his rage.

Wynter also implied that Obaysch's temper might be the result of sexual frustration and speculated that it might 'improve when his young bride in the adjoining room is presented to him.'

This was not the Obaysch of *Punch* and the *Illustrated London News*, and reading the various minutes and reports of the Zoological Society one finds very little to suggest that there was a problem. Yet Bartlett's account above suggests there was, and coheres perfectly with Wynter's.

And with the advent of Adhela the problem was doubled.

What a devastating image Wynter presents. Here is a captive animal so mad with rage and grief that it smashes itself repeatedly against solid oak and metal. This is one reason why the image of Obaysch was so carefully manipulated and controlled, and we should remember it whenever we look at captive animals.

From 1855 comes what is the archetypal image of Obaysch. This is reproduced on the cover of this book. It is a photograph taken by Don Juan Carlos María Isidro de Borbón, Comte de Montizón, pretender to the throne of Spain as Juan III and to the throne of France as Jean III. He found it hard to get work as a king so became a founder member of the Photographic Society instead. The magnificent picture shows Obaysch resting beside his pool, which reflects his somnolent bulk almost perfectly.

There are two points of significant interest in this photograph beyond the simple fact that it is a wonderful image of a fascinating animal.

The first is that the scar on his side, remnant of the injury he received when he was speared on capture, is clearly visible as a roughly circular blemish about half way down his body and nearer his back leg than his front leg (which is also consistent with accounts of the way it was gradually moving rearwards as he grew bigger).

The second is the audience. They are standing behind a railing but, one would think, still within about a yard of Obaysch's body. On the right stand two men and three children, all respectably and well dressed. To the left are a man and woman and a child. The man is respectably dressed and the woman may be wearing a uniform – is she perhaps a nanny to the child? She seems to be leaning into the man slightly, as if to talk to him or listen to him. Is this a gentleman and his well-tended child enjoying a day at the zoo? Between these two groups are two figures which are the most ambiguous. They may be wearing light summer clothes but, equally, they may be wearing working man's smocks. They have informal round hats which contrast with the smarter hats and caps of the other men in the picture. Are they working men who have saved their money to see the famous sight or are they, perhaps, zookeepers? The earliest images of Zoological Society keepers show them in dark clothes and caps but these men may be wearing smocks to protect their clothing. There is a photograph of Frank Buckland in his working gear in which he is wearing a hat very similar to the ones worn by the men in this picture. The other spectators are standing upright and holding the bar in front of them with both hands, their gaze fixed on the hippo. These two are leaning forward with their arms folded so that their elbows are resting on the bar. They are also looking intently at Obaysch but are craning forward to do so. Is this evidence that they are taking a professional interest, that they are working men who have a different code of body language from the gentlemen who surround them, or are they simply taking a more informal approach to the whole experience?

This picture may or may not represent a socially diverse group visiting Obaysch. What stands out, however, is Obaysch's magnificent

1 The Life and Times of Obaysch the Hippopotamus

indifference to having so many observers. This may be because, by this time, he was used to the crowds. It may be that he was not now alone and somewhere just over a wall he could smell and hear the infant Adhela. But it is more likely that this image shows that most of the time he simply was not interested in the people who came to see him and that in spite of all the attempts to construct him as a creature with a human-like personality he remained, like most animals, radically different and distant.

And, of course, the photograph is taken in such a way as to suggest that it is the humans who are the captives and not the hippo.

By 1858 the gaze of visitors when confronted with Obaysch was sufficiently identifiable for Edward Trelawny to refer to it to describe an incident in Greece many years earlier:

> The two chiefs at first looked at the Major's novel proceedings with curiousity as visitors in the Zoological Gardens do at the hippopotamus ...[48]

To encounter a hippo in London, to meet Obaysch, was to encounter a strangeness that was hard to credit even when it was lying in front of you, snoring. It is almost as if Obaysch imported a new way of seeing: a gaze that took in the reality of a living animal but could not connect it with anything that made sense within a familiar frame of reference. When one saw Obaysch even the conventional hierarchy by which animals were captive and humans free might be experienced in reverse. This is one reason why so much time was spent finding familiar comparators or contexts for Obaysch. If these were not available as substitutes for direct perception he would be radically incomprehensible.

In 1859 Obaysch and Adhela were briefly joined by another hippo known as Bucheet. He had been captured by John Petherick,

---

48   E.J. Trelawny, *Recollections of the Last Days of Shelley and Byron* (Cambridge: Cambridge University Press, 2011), p. 259.

who was now the British consul in Egypt and a significant collector of animals in his own right, when only a few days old and shipped over together with a number of rare birds and animals. The birds and animals were presented to the zoo but it appears that Bucheet was merely a guest. He was eventually sold on to Barnum, who exhibited him in the United States as the first live hippo to be seen there. Barnum let it be thought that he had paid a king's ransom for the animal from a local tribe. In fact, he paid Petherick £3000, which is certainly a prince's ransom. Whether Petherick's price was too steep for the Zoological Society or whether it was felt that two hippos were all that they could cope with is unclear.

It could have been worse: Bucheet was originally one of four hippos Petherick had captured. His boat capsized on the Nile as he brought them back. One was trapped in the boat and drowned, one swam to the shore and ran off, and one died 'of natural causes' before it could be transported. Bucheet was shipped in far less style than Obaysch had been: he travelled in a horse box and had water thrown over him to stop him drying out. He arrived in poor shape – his skin had gone dry and hard – but soon recovered his condition, not least because his Arab companion, Salama, cared for him. Like Obaysch's Hamet, Salama travelled with Bucheet and looked after him day and night.[49] We get a brief glimpse of Salama's stay at the zoo in this nasty little comment by Charles Dickens:

> The Zoological Society ... have got attached to the service of Prince or Princess Hippopotamus a 'native', who seems, from a

---

49 Petherick wrote up his own account of these events as *A Full and Interesting Account of the Great Hippopotamus from the White Nile* (Boston: J.H. & F.F. Farwell, 1861), which was pirated two years later by the New York publisher Booth to cash in on Bucheet's increasing fame as he was exhibited. Although Barnum was the eventual buyer, it seems that Bucheet went through other hands first. Bucheet and his trip to the USA will be encountered again in Chapter 3. The *Proceedings of the Zoological Society of London* for 1860 refers on p. 195 to the Petherick hippo. See also H. Scherren, *The Zoological Society of London* (London: Cassell & Co., 1905), p. 121.

1 The Life and Times of Obaysch the Hippopotamus

casual glance, to have all the qualities of the ape in good development. He holds out his paw with grinning cries for a halfpence to those who visit this department, and to ladies especially, grinning and staring at them in a way that very offensively carries out the resemblance that has been hinted above. Perhaps the members of this otherwise admirably conducted society are not aware of the proceedings of this one of their servants, and that a word of warning is very much required to prevent this swarthy gentleman from annoying the visitors to the Gardens.[50]

If this is Salama (and the date of the piece suggests it must be), this represents one of the only appearances in the press of this particular hippopotamus. We might also note that Salama is treated very differently to Hamet, who was almost always presented as a genial and kind friend to Obaysch, albeit one who was pathetically doomed to bask vicariously in the glow of Obaysch's celebrity. Salama is last spotted in Barnum's circus. He was advertised on 12 August 1861 as being as much worth seeing as the hippopotamus itself:

[Salama,] who is himself a curiousity, as a specimen of that historic Arabic tribe of men, who exhibits all the dignity of that Oriental race; the only man who can control or exhibit hippopotamuship, is in constant attendance.

Notice that the distinct effort to align Obaysch with the Arabic world and not with Africa is also to be seen in the case of Salama and Bucheet.

After the arrival of Adhela, Obaysch almost vanishes from sight in the press. The previously frequent mentions, the steel engravings and the satire die down. There was an alarming incident sometime

---

50 C. Dickens, *All the Year Round* 3 (1860), p. 142.

in 1860 when Henry Hunt, who was now Obaysch's keeper – although there may have been two keepers at this point, presumably to allow for some kind of shift pattern – rushed into Bartlett's office shouting the words that Bartlett must have dreaded above all:

Obaysch is out!

It seems that Obaysch had taken advantage of a temporary closure to the door of his pen, dislodged it, and was now trotting up the path with 'his huge mouth curled into a ghastly smile.' Hunt tried to tempt him to go back by offering hay but although Obaysch was all too happy to eat what he was given it wasn't enough to tempt him back. Bartlett was ever resourceful and knew that Obaysch had conceived a mighty dislike for Matthew Scott, the elephant keeper (perhaps he remembered the humiliation of being replaced in the public affection by a baby elephant). Bartlett persuaded Scott to act as a decoy. He stood and shouted at Obaysch with the desired result:

> Ugh! roared the beast, viciously, and wheeling his huge carcase suddenly round rushed with surprising swiftness after the keeper.

Bartlett may or may not have known that a hippo can easily outrun a man over a short sprint. One imagines that Scott did not know. Fortunately for Scott, he reached the pen before the hippo did and rushed out via the steps that had been built as an emergency exit for keepers who found themselves in trouble. A troupe of other keepers, led by Bartlett, had sprinted along behind. The main gate was slammed and locked and all was well again.[51] It is noteworthy

---

51 This and the following quotations regarding this incident all derive from the account given in A. Bartlett (ed. E. Bartlett), *Wild Animals in Captivity*, pp. 72–86. Buckland's account was written up by George Bompas in his *The Life of Frank Buckland* (London, Smith, Elder & Co., 1886), pp. 282–285. The phrase 'Obaysch is out' derives from Buckland's account. It sounds more likely to have been the kind of thing a panicking zookeeper might have said than what Bartlett remembered: 'Master, master, the hippopotamus is out.'

1 The Life and Times of Obaysch the Hippopotamus

that Obaysch and Adhela were not together (or they would both have been out) and that there was already sufficient doubt about Obaysch's temperament to build an escape route from his pen. Buckland also claimed that just as Obaysch was being safely locked up a reporter arrived in a cab, having heard that there was something up. But Bartlett was able to deny that Obaysch was on the loose, which is perhaps a small but apt example of the manipulation of Obaysch's image.

As far as I can see this brief trot up the path was the only time that Obaysch ever experienced anything like a break in his captivity or relief from the confines of his pen. This is worth considering when we think about him, and about any captive animal.

The account of this incident given by Bartlett differs subtly from that given by Buckland – who wasn't there. Buckland suggests that in order to persuade Scott to be the human bait, Bartlett presented him with a banknote. We must remember that zookeepers were not especially well paid, and in the 1860s English banknotes were large and magnificent objects (until 1853 they were still hand signed by the bank cashiers) that would rarely come the way of a working man. Buckland pictures Scott sizing up the hippo, mentally calculating the distance and speed he would need to run, and then contemplating the banknote before agreeing to take on the job. Bartlett denies this ever happened, and one would have to side with him, partly because he was on the spot and partly because Frank Buckland, good scientist and experienced naturalist that he was, was also a 'character' who would readily have adopted what we currently call 'alternative facts' in pursuit of a well-rounded tale.

Velvin, in *Wild Animal Celebrities*, gives a slightly different version which may or may not derive from a new source. Robert Baden-Powell, who loved to visit the hippos at the zoo and wanted one for a pet (he also loved to shoot them in the wild), gives a different version, but it reads like an enhanced version of Bartlett's account. In *Scouting for Boys* Baden-Powell included the following Zulu chant: '*Ingonyama – gonyama! Invubu! Yebo! Yebo! Invubu*', which translates as 'He is a lion! Yes! He is better than that! He is a hippopotamus!'

This is not, however, an insignificant detail. It does not merely reflect the differing characters of the actors involved. It also tells us something about how the Zoological Society was attempting to manage Obaysch's image. What Buckland's and Bartlett's accounts have in common is a tendency to treat this potentially disastrous event – after all, if they hadn't managed to get him back into his pen they would surely have had to shoot him, and that would have been one of the greatest public relations catastrophes ever to befall the zoo – as having comic potential. It seems to me that this use of humour is part of a consistent strategy to distract the reader from the brutal fact of Obaysch's aggression and to maintain the image of him as a gentle giant fully acclimatised to the genteel rhythms of English life, but with a loveably grumpy side.

Even in 1850 *Punch* was able to imagine a procession of animals bearing bound volumes to the Great Exhibition. The fourth volume was to be carried by:

> The Hippopotamus: only a just tribute to the good nature that redeems ugliness, and turns what otherwise would be a monster, to quite a pet.[52]

The image of the gentle, good hippo established at the beginning of Obaysch's captivity was to be maintained at all costs, and humour was one way of deflecting attention from the more worrying aspects of his behaviour. Towards the end of the century the naturalist and explorer Mary Kingsley, who encountered hippos in the wild so knew only too well what they were like, was defusing their aggression with humour in her immensely popular and very amusing *Travels in West Africa*.[53] Even today, there is an expectation that hippos will be comic; this appears to have grown out of how Obaysch was seen and depicted in the 1850s and beyond.

---

52   *Punch* XIX (1850).
53   M. Kingsley, *Travels in West Africa* (London: Macmillan & Co. Ltd, 1897).

1 The Life and Times of Obaysch the Hippopotamus

One wonders if Obaysch's lowered media profile after 1853 was actually due to his increasingly difficult temperament. Andrew Wynter's description of his terrible rages comes from 1855 and Adhela, whose aggression the Zoological Society was more prepared to acknowledge, was always seen as tricky to manage. We might speculate that by 1854 the Zoological Society had on its hands two angry and difficult animals who were hard to manage and expensive to keep. Yet it was on these animals that a large chunk of their income depended. The society faced a strikingly modern media management challenge, and their solution was to restrict and control coverage and to massage Obaysch's image as far as possible.

Hunt also had a near miss with Obaysch when, on a very hot August night – there is no date recorded, but August was exceptionally hot in 1857 and 1868; by either year, Obaysch would have been fully grown – Hunt came back from the pub and before turning in decided he'd have a swim in Obaysch's pool (like Hamet before him, Hunt lived in the hippopotamus house; there were also quarters for six keepers in the basement of the giraffe house). What he didn't know was that a night watchman had reported that the heat was making Obaysch restless and so, on Bartlett's orders, Obaysch had been let out into the yard. When Hunt dived in he found himself on Obaysch's back (Obaysch had been standing peaceably on the bottom of the pool). Hunt quickly jumped out again. Barlett recalled of this incident that 'had the brute got at him only his mangled remains would have been found to tell the tale.'

It was not until nine years after Obaysch's death, however, that Bartlett told what is surely the worst and most revealing story of the systematic concealment of his aggression:

> Well, one day a stray dog strolled casually into the gardens and stopped before the rails of the hippo's outside enclosure. The day was warm, the pool was tempting, so the dog wriggled through the rails and sprung into the water to his doom. The hippo rose

to the surface, and roaring took the dog into his great jaws, scrunching him up to bits, which he disgorged.⁵⁴

A dog eaten by a hippo in the middle of London would surely be a difficult thing to conceal, but throughout Obaysch's life no word of this appears in any newspaper or magazine or published memoir.

In 1865 we get a glimpse of Obaysch and Adhela in a piece written by 'Peter Possum' recounting a visit to the zoo and published in the *Sydney Morning Herald* on 18 August:

> in the hippopotamus tanks Adhela and her monstrous husband were lying motionless – black masses that looked like a squadron's beef in pickle.

The years had not made the hippos any livelier, but it is interesting to note that, at this point, they may have been sharing at least one of the hippo pools. A decade later, the Shah of Persia noted that 'the hippopotamus is a wonderful thing' when a visit to the Zoological Gardens was included as part of his visit to England. This shows, if nothing else, that hippos were still of sufficient rarity to enthrall even such a wealthy and eminent tourist.

Like Adhela, Obaysch managed to break a tooth. This is perhaps more evidence that his behaviour may have been more aggressive than the Zoological Society cared to acknowledge. Bartlett determined to remove the stump. He had an oak fence built between the pool and the iron railing and, armed with a huge pair of forceps, enticed Obaysch towards him and grabbed the remnant tooth with all his might. Obaysch promptly tore the forceps from his hands and charged him. Safe behind his rampart Bartlett managed to recover

---

54 'Hippos' Habits', *Pall Mall Gazette*, 8 January 1887.

1 The Life and Times of Obaysch the Hippopotamus

the forceps and grabbed the tooth again, with the same results as before. On the third trial Bartlett pulled the tooth. The importance of this surgery was stressed in an account of it given by Frank Buckland. In his *Curiousities of Natural History* he makes the crucial point that Bartlett's own narrative shies away from:

> Not so long ago, the old male hippopotamus at the Zoological Gardens suffered much from a decayed tooth. In former times he would have been shot, as was poor 'Chunee,' the elephant at Exeter 'Change.

In 1826, Chunee, a much loved animal, became unmanageable when he went into musth, a periodical surge in hormones that can lead to aggressive behaviour. He was shot and bayoneted to death over a shockingly long period on the upper floor of a building in the middle of London, and his death witnessed by dozens of other caged animals. This botched execution was one of the most scandalous events of the early nineteenth century. The fact that Buckland naturally refers to it in this anecdote shows just how traumatic the very thought of it was still for all involved in the keeping of exotic animals nearly fifty years later. Had it been necessary to shoot Obaysch, even had the killing not been as brutal as the attack on Chunee, massive public outrage would have been visited on the Zoological Society. In attempting to remove the tooth Bartlett was thus doing more than some improvised veterinary dentistry: he was probably saving Obaysch's life. I think it possible to think well enough of him to ignore the glib conclusion that uppermost in his mind was the reputation of the Zoological Society and the vast revenue at stake should visitor numbers drop precipitously. Instead we should consider that he acted, at no small risk to himself, with Obaysch's welfare in mind.

Obaysch was now left to grow and to make the best of an enclosed life that required him to be awake during the day. Even so, the constant public attention, coupled with the constant stress of incarceration, was clearly an inhibition to breeding. Adhela was popular too, and in 1868 her pen had to be enlarged because of the inconvenience of a narrow passage 'on crowded days', as the *Report of the Council of the Zoological Society* put it. Given normal maturity and life cycle patterns we would have expected Obaysch and Adhela (or 'Dil', as she was known) to have begun to mate in 1858 or 1859. However, their keepers probably didn't know that hippos mate in water with the female almost totally submerged. Given the size of the hippo pool there was barely room for mating to take place and this may well account for the time it took for breeding to commence.

It was not until 1870 that Adhela commenced her first eight-month pregnancy. She gave birth to a baby on 21 February 1871. Obaysch had thus first mated a good ten years after what might have been expected from his natural life cycle. The reasons for this may also be found perhaps in the following letter published in the *Times* on 23 February:

> Both mama and baby were doing well last evening, indeed, better than could be expected, as the female is a surly brute, and the sages hommes feared that she would bite her infant in two as soon as it made its appearance. But, on the contrary, she is very affectionate towards it and, as she has a splendid udder of milk, there is every chance of rearing the calf. Of course, the public are not as yet admitted to see it, for the dam is very suspicious and jealous. Even the keepers are obliged to act on the sly for the present.

Whereas Obaysch had proved sufficiently compliant or been sufficiently cowed to be easily managed in his first couple of years, it may be that Adhela never did settle down quite as well. She and Obaysch therefore inhabited an edgy and dangerous world.

In addition, instinct is one thing, but, in hippo culture, breeding is a privilege of hierarchy. Could Obaysch be the alpha in a bloat of one? How could he know the protocols of breeding? A photograph

1 The Life and Times of Obaysch the Hippopotamus

(one of a series by Frederick York) shows Adhela and Obaysch together in the same pen in 1870, just at the time when they must have started to mate. They are standing apart and looking in opposite directions. They look much like members of a royal family do when their marriage has disintegrated but they still have to make public appearances. It is also noteworthy that they appear to be roughly the same size, even when you allow for the effect of foreshortening in the composition. They shouldn't have been. This adds to my sense that Obaysch had not developed properly under the zoo's care.

Frank Buckland described the new baby as 'about four feet long and … about fifteen inches high when standing up'. The baby was active:

> Every now and then it lifts up his head and looks stupidly about; it often wags violently its little rudder-like tail, without rising from the straw; it also shakes its ears with the curious jerking motion peculiar to the hippopotamus. I fancy this motion is to throw out any water than may have got into the ears. Mr Bartlett tells me he has already heard it answer its father's call but it does not do this often.[55]

Buckland also noted that:

> The mother has chosen an excellent place for her child, for the corner in which the little thing is lying is formed by the junction of the hot-water pipes and so that this is about the warmest place in the den.

February in London can be pretty miserable at the best of times, so while Buckland was right to note Adhela's efforts to keep the baby warm he didn't note, because he didn't yet know, that the best place for a baby hippo is in the water and that lying on the dry land is

---

55 F. Buckland, 'The Baby Hippopotamus', *Graphic*, 4 March 1871.

a very unnatural thing for hippo mothers and babies. Hippo babies are suckled on land when the mother has to feed but they are born in water and, mainly, suckle under water.

Obaysch was clearly taking an interest but what his vocalisations could have meant ... who knows?

In spite of the care lavished on the new baby and Adhela's motherly attention, it died 'from exhaustion' after only three days. It simply had 'neither the power nor will to suck.' Another commentator wondered if the cold climate was the cause:

> The little hippopotamus, whose advent caused such a flutter of excitement in the Zoological Gardens, was the first of his race born in England, but unfortunately did not live long enough to enjoy that distinction. The tropical little strangers who are born amongst us never do; the climate is too severe for their tender constitutions and they find hot-water pipes and blankets but sorry substitutes for burning suns.[56]

In fact, 1871 was a very cold year and the average temperature in London from December 1870 to February 1871 was just 2.4 degrees Celsius.[57]

Although the death of the baby was a disappointment, Obaysch's keepers had learned that he and Adhela could and would mate successfully. They also had a dead hippo to anatomise and a satirical cartoon, entitled 'The Baby Hippopotamus, Art and literature were largely represented at the post-mortem', in *Fun* (11 March 1871) shows a rather ghastly crew crowded round the body, sketching and taking notes while on the wall behind them is a picture of an adult hippo with a smile on his face gamboling under a bright sun. Beneath this scene is another image entitled 'But these were

---

56   Quoted in 'Death of an "Interesting Stranger"', *Fife Herald, Kinross, Strathearn and Clackmannanshire Advertiser*, 2 March 1871.
57   Historical weather data is taken from 'Weather in History', http://bit.ly/2J3kDYf.

1 The Life and Times of Obaysch the Hippopotamus

the real mourners', which shows the dead baby with a full grown hippo (presumably Adhela) standing over it, eyes streaming with tears, and a top-hatted man (presumably Bartlett) covering his face with a handkerchief. A theory was developed that the baby may have died of a liver defect. The post mortem showed that it was a male. Buckland made two casts of the body, which was then stuffed. One of the casts and the taxidermic specimen were exhibited in the giraffe house. This enabled people who had seen Obaysch and Adhela (who lived next door) easily to compare them with a baby specimen but was, presumably, out of their sight. The other cast was exhibited in Buckland's own 'Fish Museum'.[58]

In the wild, hippos do not usually ovulate after pregnancy for seventeen months, so it is very surprising that almost immediately after the death of the first baby Adhela became pregnant again. She gave birth at three in the morning of 7 January 1872 to another healthy-looking child. Frank Buckland went to see it and described the scene as:

---

58 *Graphic*, 4 March 1871. Frank Buckland was very concerned with fisheries and sat on various commissions looking into the fishing industry. In 1865 he founded the Museum of Economic Fish Culture in London, which consisted mostly of plaster casts of fish made by Buckland himself, some of which still exist and may be seen in the Scottish Fisheries Museum at Anstruther. See G.H.O. Burgess, *The Curious World of Frank Buckland* (London: Horizon, 1967) and R. Girling, *The Man Who Ate the Zoo* (London: Random House, 2016) for entertaining accounts of his life. *Fun*, on 17 March 1866, published a cartoon entitled 'The History of a Hippopotamus', showing a live hippo in Africa, the same animal dead and being crudely stuffed, and then the finished product: a moth-eaten and battered specimen, 'as he appears at the British Museum'. The plaster cast of the baby hippo was mentioned by Mark Twain in the journal he kept of his trip to England in 1872; see *Mark Twain's Letters, Vol. 5*, edited by E.M. Branch et al. (Berkeley: University of California Press, 1997), p. 586. It was still on show as late as 1960, when the hippo house was demolished. A haunting image of the cast in storage can be seen on the website of the library of the Zoological Society of London: http://bit.ly/2yHqBcx.

very much the same as that which I described when the last baby hippopotamus was born. The mother lay in the corner farthest away from the window, the young one lay close to her; the nose of the mother was close to the nose of the infant. Everything was painfully quiet, and the only sound was the chirping of the sparrows; the sparrows seemed to chirp louder in the hippopotamus house than anywhere else.[59]

But already it had been noted that the baby was not suckling. They had put it in the water and 'the young could swim as well as its mother'. The zookeepers had tried to take the baby as soon as it was born but Adhela was too aggressive in its defence. They had driven Adhela into her pond and then slammed the door and tried to feed the baby by hand. They even had got goats on hand in case Adhela's visibly ample supply of milk should fail. But the baby simply couldn't find the teat and Adhela seemed unable to show it the right way – the baby sucked at her ears, her mouth and her feet.

There was still hope, however. In contrast with the birth of the first baby, when Adhela had been 'restless and suspicious and would not let her child to go out of her reach', she now 'allowed her baby to walk about her cage, which her calf has repeatedly done.' However, with the lack of nourishment the baby soon lost strength and by Tuesday 9 January 'it was lying very quiet under its dam's chest, occasionally twitching its ears'. Adhela, who had now twice been thoroughly frightened and distressed by attempts to remove the baby, 'had so arranged herself as to command with one eye one of the doors leading into the cage, and with the other eye, the other door.' Even so they managed to drive Adhela off by squirting her

---

59 'Birth and Death of a Hippopotamus', *Morning Post*, 15 January 1872. In this piece Buckland calls the baby Umzimvooboo, which he claims is the African name for a hippo, but there is no evidence anywhere else that this was a name given to this hippo. Other quotations in this section are from 'The Young Hippopotamus', a letter published in the *Times* on 10 January 1872 before the baby died and before the successful attempt to remove her from the pen.

1 The Life and Times of Obaysch the Hippopotamus

in the face with water from a high-pressure pump at about midday on the Wednesday. They took the baby to the other end of the gardens where Adhela could not hear it; Bartlett seriously believed that should she wish she could demolish either the bars or the brick walls of her pen. Bartlett noted that the baby was very cold (1872 was, in fact, the wettest year on record to that time and would not be surpassed until the truly miserable 2000–2001, so January would have been chilly) and wrapped it in blankets and cotton wool. This had the desired effect: the baby warmed up and it was then offered a feed of ass's milk. It at first refused and would not take anything while it could see or hear humans, but when it was blindfolded and everyone kept quiet it started to suck, took about a pint, and then fell asleep. Over the next six hours it took two more pints but it was too little, too late and round about seven o'clock in the evening it died. The next day a very full account of its brief life and sad end was published in the *Daily Telegraph*.

By Thursday morning it had already been dissected and sketched for the *Illustrated London News*. It was five feet nine inches nose to tail. The remnants were to be sent to Oxford for further study. The post mortem showed it was female. *Punch* (27 January 1872) commented that:

> Of all the odd kinds of consolations under affliction, the last suggestion seems to Mr. Punch the oddest. We are mourning the demise of the no-horned infant Hippopotamus in the Regent's Park, and we are told to be cheerful because a two-horned Infant Rhinoceros has gone to Madrid. The doctrine of compensations was never pushed much further even in a Scottish sermon.

All this time Obaysch was somewhere close. He is not mentioned in any of the accounts of the birth and death of his second child. But he must have been in an adjacent pen. He would have heard Adhela's grunts during and after the birth, her enraged cries when the first attempt was made to remove the newborn baby and twice more at the failed second and successful third attempts to remove it. He would have heard the cries of distress of the baby itself when it was removed. But no one mentions him and he remains a brooding and, one imagines, distressed and angry presence on the outside of this little pachydermous tragedy. He is an alpha who has lost control of his bloat and not at all a Victorian *paterfamilias* insouciantly puffing at a Lonsdale and looking discretely at *The Pearl* while the maids rush up and down stairs with boiling water.

Once again, seemingly against the normal cycle of hippo fertility and reproduction, Adhela was soon pregnant for a third time. On 5 November 1872, only ten months after the death of the second baby, Guy Fawkes appeared. Almost a year later it emerged that Guy Fawkes was actually female. But by that time it was too late to change her name and Miss Guy Fawkes was the best that anyone could come up with.

The experience of losing the two previous babies had caused Bartlett to think about his options, and before Guy Fawkes' arrival he had made some modifications to the hippo house. In 1872 he laid on a water main dedicated solely to the hippo pool to enable it to be filled more quickly; the same year the hippopotamus house was enlarged, an expansion described in the *Report of the Council of the Zoological Society* as being made 'in anticipation of the birth of another young one.' In 1871 a modification described in that year's *Report* as designed 'to allow the young one to be easily removed and brought up by hand' clearly had not worked. Bartlett had also received a number of letters advising him on how to do better – some of them sensible, some of them cranky. One, from 'M.W.', pointed out that hippos suckle under water, another from 'A Friend to Animals' pointed out that nursing animals need peace and quiet and that hippos were probably no exception. The keepers at Amsterdam (where they had lost five baby hippos, so why Bartlett took their advice, who knows?) had told him that under no circumstances should the baby be allowed in the water. He had seen that the second baby had swum and that although it had died it

1 The Life and Times of Obaysch the Hippopotamus

seemed to have survived its dip, so he determined on an alternative approach and let Adhela take the baby into the pool with her. He also ensured that the pool was, thanks to 'Mr W. Hill's Improved Flue Boiler and Furnace', heated to approximately the temperature of the White Nile. By doing so he discovered that what M.W. had said was true and that young hippos do suckle under water. The zoo had again taken the precaution of bringing in goats and asses to feed the baby if need be, but this was not to be necessary. In addition, Bartlett had seen how anxious Adhela could get when she had a young one so he instituted a regime of silence in the hippo house and minimised movement where Adhela could see or hear it.[60]

As one press report put it:

> Madame Hippotama, whose temper was never good at the best, is not to be trusted even by the keepers, who think that if she were disturbed by the vulgar gaze of the multitude, she might neglect or possibly trample on her calf ...[61]

In fact, Bartlett later reminisced that at this time Adhela was so aggressive that:

> I was afraid she would kill herself and her baby as well. We were compelled to feed them through the ventilator and never dared go into the house.[62]

Characteristically, Frank Buckland saw it more sympathetically:

> If she were once to be put out the poor old thing, who looks exhausted and anxious, would probably, in her alarm, get up,

---

60 Bartlett (13 November 1872) chose to report on the week's events reasonably accurately but as a series of comic bulletins by L. Ephant and Rye Noserus.
61 'Hippos' Habits', *Pall Mall Gazette*, 8 January 1887.
62 F. Buckland, 'Birth of a Young Hippopotamus', *Morning Post*, 6 November 1872.

rush about, and possibly not suckle her child, or else trample by accident upon it.[63]

This encouraging start meant that Guy Fawkes was allowed to swim to her heart's content. Bartlett became so nervous when she seemed to be submerged for an unusual length of time (Guy was only eight days old at the time) that he gave orders for the pool to be drained – there was what was essentially a huge bath plug at the bottom – but then Guy bobbed to the surface, having been under for longer than Bartlett thought possible. Frank Buckland recorded this event and his account of it shows not only how precious the new baby was but also the especially charming nature of Buckland's approach to zoology and the emotional intimacy that permeates his engagement with the many animals that crossed his path:

> A few days after the birth of the young one, Mr Bartlett was watching it swimming about the tank. It then suddenly dived, but did not reappear for such a long time that he thought it had had a fit and was lying drowned at the bottom of the tank. He therefore made arrangements to have the large plug pulled out – this plug had been fixed expressly for this purpose – and to run off the tank quickly so as to resuscitate the little beast if possible. They were just going to do this, when Master 'Guy Fawkes' suddenly reappeared, shaking his funny little horse-like ears, with a hippopotamic grin on his face, as much as to say, 'Don't be frightened, I am all right; you don't know all about me yet!' The little beast had remained, without blowing or taking breath, actually under water for nearly twenty minutes. The parents have never been known to be under much over three minutes.[64]

---

63 Buckland also gave a charming account of this event and other aspects of 'Dear little Guy Fawkes' in a piece in *Land and Water* which was republished in the *Blackburn Standard* on 20 November 1872.
64 F. Buckland. 'The Young Hippopotamus', *Times*, 8 November 1873.

1 The Life and Times of Obaysch the Hippopotamus

Whether or not Guy was really under water for twenty minutes (and it seems unlikely) this passage shows the biggest problem that Bartlett faced in trying to keep his hippos healthy and, in the case of the babies, alive at all: he really didn't know much about their behaviour (no one did) and had only anecdote, travellers' tales and the somewhat risky advice of his Dutch colleagues to go on.

The care of Guy seems to have involved allowing the baby and Adhela to be together in the pool much more, as they would have been in the wild. Generally, this regime (being much more attuned to the hippos' natural inclinations) enabled Guy to thrive and Adhela to calm down, although on one occasion Guy got into difficulties when she couldn't get out of the pool and Adhela was unable to help. Bartlett and the keepers had to watch helplessly until the little hippo managed to struggle from the water on its own. Note that even though they risked losing their prized baby hippo, the thought of encountering an angry Adhela was sufficient to keep them outside the pen.

When Buckland noticed that Adhela 'manifested no anxiety at the presence of such a number of strangers' he was referring to the initial private viewing offered to the fellows of the Zoological Society on 23 November, which was allowed when, by 21 November, Adhela 'was becoming less savage and excitable when approached.' In fact, this viewing seems to have been open to the general public. The hippopotamus house was re-opened for the general public from 12 p.m. until 4 p.m. and at 3 p.m. Adhela was fed. In keeping with the need to provide a show, Adhela had been kept slightly short of food so that she (and therefore Guy) would come out of the pool the moment the food arrived. The next day neither Adhela nor Guy were comfortable with the crowd, which as the *Penny Illustrated Paper and Illustrated Times* (30 November 1872) said, 'appeared distasteful to both mother and baby', who spent most of their time submerged in the water. For all the excitement Guy Fawkes thrived and after a month the *Illustrated London News* (7 December 1872) reported that she had grown to be five feet (1.52 metres) long and to weigh two hundredweight (102 kilograms).

While we had only a shadowy sense of Obaysch's presence at the first two births, we glimpse him a little more clearly this time:

> The old father in the next den talks to his wife and child by means of sonorous gruntings, and they answer him.[65]

Even so, although Obaysch appears at this point to be taking a gentle paternal interest in proceedings, it was felt politic not to let him near Guy Fawkes for another eight months. This may not have been so much because he was seen as a threat to the baby (although he was described as the 'disagreeable old father') so much because of the risk of spooking Adhela. The caution proved to be well founded:

> Obesh [sic] was gently munching his breakfast of grass in the outside den, when, at a given signal, the portcullis of the mother's den was gradually raised, and the two heads appeared, gazing out with a most comical expression. Seeing his wife, the old man left off munching his grass, grinned a ghastly grin, and loudly trumpeted 'Ump,' 'Ump,' 'Umph.' Little Guy Fawkes then came out from behind his mother, with the action and stiffness of a pointer when he has discovered a covey of birds; gradually and slowly he went up to his father, and their outstretched noses were just touching when the old woman sounded the signal for war, and rushing past the young one, fairly challenged her lord and master to single combat. He instantly retreated a step or two, and his wife began to pretend to munch at the grass, keeping her eyes fixed spitefully upon him … After gazing at each other for about a minute, old Dil – for that is the female's name – made a savage rush at her husband, and simultaneously both animals reared up right on their hind legs, like bull dogs fighting. They gaped wide

---

65 F. Buckland. 'The Young Hippopotamus', *Times*, 8 November 1873.

1 The Life and Times of Obaysch the Hippopotamus

their gigantic mouths, and bit, and struck, and lunged at each other savagely, while the grass fell out of their great coal-scuttle mouths onto the battle field ... When they had settled on their four legs again, the old woman followed up her advantage by giving her husband a tremendous push, 'well hit,' with her head; and while the cowardly old fellow sneaked backwards into his pond his wife trumpeted a triumphant signal of victory from the bank.[66]

After this Guy Fawkes crept out and Adhela, with Guy Fawkes perched on her back as she might have been in the wild, went down into the pool but threatened Obaysch if he moved from the end where she had him pinned. He made 'one attempt to get out of his corner, and retreat into his den, but his artful old "missis" was too quick for him, cut off his retreat and drove him back.' After a while they settled down and Adhela started to hiss:

> When the keeper heard this he said, 'They are all right now, sir; they'll not fight any more. See, the old man's beginning to smile and he has uncocked his ears and left off staring.'[67]

The fuller significance of this incident will be discussed in Chapter 2, where the various meanings attributed to Obaysch will be explored. However, in purely biographical terms, we might suppose certain things. First, as other reports suggest, Adhela may have been more aggressive than Obaysch, and the decision to keep them in separate pens may have been more than a logistical one. Second, a fully grown adult male hippo would not normally be defeated by a female one, even if she is a mother defending her baby. By this time Obaysch could have been more than double the weight of Adhela. If he was, then although he could still be damaged by a bite from

---

66 'A Hippopotamus Fight in the Zoological Gardens', *Derby Mercury*, 30 July 1873.
67 'A Hippopotamus Fight in the Zoological Gardens', *Derby Mercury*, 30 July 1873.

her, he should have been invincible in the sumo-like contest that is described above. This incident surely shows that something had gone wrong with his normal development, and that he was nothing like the size and weight that would usually be attained by a twenty-year-old male hippo. Was Obaysch in poor health? Did he deserve to be called a coward?

This episode also shows that the keepers (by this time Michael Prescot and Arthur Thomson, who were each given the Zoological Society's bronze medal – Bartlett picked up the silver – and a bonus for their work in bringing Guy Fawkes to healthy infanthood) had sufficient experience to know the difference between aggressive and peaceful hippo behaviours, which suggests that they had seen plenty of both sorts.[68]

Buckland visited Guy Fawkes on her first birthday and recorded his visit in the *Times* (8 November 1873). He noted that Prescot was able to entice Adhela and Guy Fawkes out of the water and that Obaysch's face was much longer and sharper, with more prominent eyes, than Adhela's. He also tells us that Bartlett believed that another hippo would be born in April. However, there is no record of a birth in April and no record of a miscarriage. It may be that Adhela and Obaysch had been seen mating again, but clearly nothing came of it. Buckland also mentioned that Barnum had said he would buy the new baby, and then adds a comment which suggests that the whole thing may be a joke: 'I doubt if he will; let him go and catch a wild baby hippo for himself.'

---

68   Hippo keepers have a good record of winning Zoological Society medals. In 1927 Ernie Bowman, known as 'The Professor', was awarded a bronze 'for services rendered in the successful rearing of a young hippopotamus 1926–27'. This was a male called Jimmy. Bowman had previously been involved in the care of the pygmy hippos and there is a wonderful photograph by F.W. Bond showing him feeding one milk from a baby's bottle and another, taken in 1923, showing him with the hippos Bobby and Joan (http://bit.ly/ZSLBowman). Buckland referred to Prescott as Adhela's and Guy Fawkes's 'personal valet'.

1 The Life and Times of Obaysch the Hippopotamus

After the birth of Guy Fawkes the hippos become much less prominent in the media and Obaysch sank into a gentle retirement. By 1872 he was only twenty-three years old, less than half way through a reasonable hippo lifespan, and yet he is always referred to as old. Certainly, after the birth of Guy Fawkes there seems to have been no more mating, which seems to have been as much due to a lack of inclination on Obaysch's and Adhela's parts as to a decision of the Zoological Society to keep them separate. The generally unsocial nature of hippos in the wild is varied only by what appears to be a tendency for mothers and daughters to form a bond. Arguably, Obaysch was now even further isolated, as Adhela and Guy Fawkes became a social group from which he was excluded both by his gender and by the iron bars and brick walls that separated them for much of the time.

Obaysch still had one further challenge to face. In 1877 another hippo arrived:

> The original pair of hippos, obtained from the Viceroy of Egypt (the male in 1851, the female in 1854), being both now well advanced in years and having ceased to breed, the Council thought that it would not be right to miss an opportunity of obtaining a mate for the society's young female [i.e. Guy Fawkes] both in the Garden on the 5th of November 1872.[69]

The new hippo was purchased from Amsterdam Zoo at a cost of £800. He was born on 3 August 1876 and arrived in London together with Mr Hegt, a Dutch zookeeper, on 20 June 1877. This hippo is a shadowy figure. He does not feature in the press, he did not manage to mate with Guy Fawkes, and he died in 1886 at the very young

---

69   *Proceedings of the Zoological Society of London*, 1877, p. 680.

age of ten. Given his hardly noticeable presence in the zoo one can only assume that it was sickly. He may have been called Anthony. Certainly it is odd that more wasn't made of what was, after all, still a rare and expensive animal purchased in pursuit of what we would now call a captive breeding program.

Obaysch would have known that a new young male was in the vicinity. Did he require deferential behaviour when they caught sight of each other? There is simply no record of any encounter between the two. But Obaysch passed away on 11 March 1878, only eleven months after the new animal arrived. He had not been well for a couple of years and during the winters he was 'observed to be in an unsatisfactory state of health, getting thin and emaciated', although his condition improved in the summer months when more green food was available.[70] This comment makes me think that Obaysch was always out of condition and simply could not derive the nourishment he required from the available diet. It would surely explain why it took so long for him to mate with Adhela. I also wonder whether a comment made about Adhela in 1855 – 'She has continued to grow as rapidly as the male when at the same age, and has never had a day's illness'[71] – implies that Obaysch may always have been slightly sickly and hadn't thrived. Or could his broken tooth, the result of his captivity, which had caused him to bite a material far harder than anything found in a hippo's natural environment, have caused a fatal inhibition in his ability to eat or a chronic infection that fatally undermined his health over several years?

His passing was noticed in the press, with the following as a typical example:

> The hippopotamus at the Zoological Gardens died on Monday night. He was caught while quite a baby in 1849, on the Island

---

70 Bartlett, *Wild Animals in Captivity*.
71 *Report of the Council of the Zoological Society*, 1855, p. 72

1 The Life and Times of Obaysch the Hippopotamus

of Obaysch, on the White Nile, and created much excitement on his arrival at the 'Zoo' in 1850. Down to the time of his death he continued to be a great favourite with the public, the arrival of his more juvenile mate 'Adhela' in 1853 [sic], having in no degree lessened his attractiveness.[72]

The *Illustrated London News* (16 March 1878) commented, in a very brief notice, that Obaysch 'created immense excitement at the "Zoo"' in 1850.

The Zoological Society believed he had died of old age, and Bartlett erroneously concluded that hippos do not live to a great age. A postmortem was performed and no organic illness was found. In 1879 the society was treated to a learned paper on Obaysch's brain, the size of his stomach (eleven feet long, or 3.35 metres), his liver, and his small gall bladder. Although it finds no obvious illness, this report does confirm that all had not been well with Obaysch for a number of years. We learn that he was suffering from leg ulcers 'which were much inflamed during the winter, then the summer season.' We learn that because 'the coldness of its tank did not allow of its remaining in the water for any length of time' his skin was dry and cracked. We learn that he was twelve feet (3.66 metres) long. This is at the lower end of the average range for a mature hippo (ten to sixteen feet), which may account for his defeat in the fight with Adhela, who could, if she had been even in the middle of the average size for a female hippo (nine to fourteen feet), have been as big as him or even bigger.[73] So Obaysch was not well, had been in bad shape for years, and had not grown to anything like his potential full size. Just as his aggression was kept as much a secret as possible, so we must conclude that his failure to thrive and the visible marks of a

---

72 *Western Times*, 14 March 1878.
73 A.H. Garrod, 'On the Brain and Other Parts of the Hippopotamus (*H. amphibious*)', *The Transactions of the Zoological Society of London* 11 (1880), pp. 11–18.

lack of care, in spite of the fact that he was undoubtedly much loved, were similarly suppressed until the very end.

There will be some discussion about the various poems written about the London hippos in Chapter 2, but for now it is worth quoting a few lines from the obituary for Obaysch that appeared in *Punch* on 23 March 1878, which imagines Obaysch's last thoughts as he lay dying:

> Well, I have had my day,
> Better indeed had men but let me stay
> In sedgy Obaysch, island of my birth,
> That cosy lair on White Nile, whence white men
> Brought me, a babe, to this close tank and pen.
> I dreamt of it last night – the unctuous ooze,
> Where one might take one's ease, and bask and snooze,
> The warm Egyptian glow, the wap and wash
> Of water in the reeds! Once more to dash
> Big-bulked through rushy reaches, strong and free!
> Methinks 'twould yet revive me. But I see
> Kind BARTLETT'S boding head-shake. Good old man!
> He has done all he can
> To make my cage a home for a poor brute,
> If in this clammy clime one could strike root.
> Ah, well! I've had my triumphs and am yet
> A Public Pet!

This is, perhaps, the only acknowledgement that Obaysch was a wild animal with a wild life and that being in a zoo in the middle of London was imprisonment pure and simple. This belated recognition of Obaysch's agency may be a function of differences in public sentiment and attitudes to animals between 1850 and 1878 but, if it is not, it represents a touching farewell to a creature who had been part of London society (and, for a while, a mainstay of *Punch* cartoons) and a familiar sight for over a quarter of a century.

1 The Life and Times of Obaysch the Hippopotamus

Obaysch was gone. Adhela lived on for another few years and passed away on 16 December 1883 at the age of about thirty. It was commented that:

> It is thus evident that about 30 years is the limit of hippopotamine existence, as it is not likely, judging from the state of the teeth and bones, that either of these animals would have been able to support existence so long in its native wilds as under the favourable circumstances in which it lived in the Regent's Park.[74]

In fact, they were each robbed of nearly half their natural lifespan. Both Obaysch and Adhela died young. Recently the hippo Bresta died at Kiev Zoo, aged fifty-eight; Lotus, an American circus hippo, lived from 1903 to 1954; Pete lived in New York Zoo until he was fifty; Targa, a hippo born in Leipzig Zoo in 1934 died in Munich Zoo as recently as 1995 at the age of sixty-one years and two months.

Adhela stimulated at least two verse obituaries, neither as touching as the one for Obaysch quoted above. In *Punch* the poem concentrates on Adhela's value as an attraction to the zoo and ends:

> She's gone. Let's shed a tear, and thus
> Lament our Hippopotamus
> Hic jacet, 'neath a tumulus,
> Adhela Hippopotamus!

In *Moonshine* the treatment is even more lightly humorous:

> Farewell, Adhela, sweet and gentle friend,
> No more shall I your graceful form behold,
> No hours beside your pond pellucid spend
> And bravely risk the chance of catching cold.
> No; you are gone, and I should ne'er again

---

74   *Cornishman*, 18 January 1883.

> Unselfishly regale thee with a bun,
> The only palliation of my pain
> Consists in gazing at your little 'un.

The difference between Obaysch's and Adhela's mock obituaries is marked and can only really be accounted for by gendering, partly in the deliberate service of promoting Obaysch and partly as an almost inevitable consequence of habits of mind about male and female characteristics. Adhela simply did not inspire the same affection or imaginative engagement as Obaysch. Her construction as a difficult animal, and specifically a more difficult animal than Obaysch, which I have argued characterised the Zoological Society's attitude towards her throughout her life, comes out even in these ephemeral responses to her death.

Guy Fawkes carried on until 20 March 1908. There were some images of her in the *Graphic* (10 September 1892) showing her asleep, yawning ('The hippopotamus makes an urgent appeal for buns', the caption read; the obituary in *Moonshine* quoted above suggests that throwing the hippo a bun may have been a common activity) and being fed – a keeper hands her fistfuls of hay from a wheelbarrow. But otherwise, she is not very visible in the record, although there is a marvellous photograph of her looking not entirely happily at 'the giraffe house cat', which has got into her terribly sparse pen. There is no vegetation, nothing soft, just hard concrete and brick surfaces surrounded by iron bars. This is, of course, the only world she knew. In 1887 Bartlett said of her:

> It may surprise you to know that our hippo (and all hippos) is a very intelligent animal. She hates the sight of the shabby clothes of a workman, but is unmoved by the garments of a lady or a gentleman. But, though intelligent, she is very sulky, and often spends a whole day in the tank.[75]

1 The Life and Times of Obaysch the Hippopotamus

Interestingly, although a long exposure to hippos had taught Bartlett that they were not stupid, as was frequently said of Obaysch, it had not taught him that their natural habit was to spend the day in water, or that this was not necessarily indicative of a bad mood. It is almost as if, since the very first day of Obaysch's arrival, there was a general disappointment that the hippos weren't more lively.

A brief account of Guy Fawkes written in 1894 continued the theme that she was a bad tempered and somewhat peevish creature and suggested that she took after her mother 'when she is bent on mischief and goes "on the rampage" to some purpose.'[76] This last phrase is another interesting example of how presentation of hippopotamus aggression was gendered, for it was surely derived from Dickens' *Great Expectations*, in which 'on the rampage' is used of the dangerously unpredictable anger of Mrs Joe Gargery. Guy Fawkes was said only to like one keeper at a time and even chased a keeper she didn't know out of the pen, giving him a bite as he swung himself over the fence. The fact that he didn't lose a leg is remarkable.

She was also, like Obaysch, a bit of a disappointment in the flesh. In 1894 the pioneering social realist writer Arthur Morrison took a break from describing the violent currents of life in the slums of the East End of London and between 1892 and 1894 published a series of humorous illustrated articles, 'Zig-Zags at the Zoo', in the *Strand* magazine. In 1893 he described a visit to see Guy Fawkes in a 'Zig-Zag Pachydermatous':

> Still it cannot be too widely known that the hippopotamus does move sometimes, and some insignificant proportion of the visitors (about ¼ in 10,000, I believe) witness the feat. But even then she rarely does more than change her elevations – just brings her north elevation south, for a change of air. It is a grave

---

75 Bartlett, 'Hippo's Habits'.
76 'Miss Guy Fawkes of the Zoo', *Monthly Packet*, 1 February 1894.

and solemn rite this turning about, and it proceeds with properly impressive deliberation. She rises by a mysterious process, in which legs seem to take no part; she anchors her face against the ground, as regarding her head in the light of a great weight (which it is) dumped down to prevent the rest of her being blown away by an unexpected zephyr. Then, with her weighty muzzle as a pivot and centre, she executes a semi-circular manoeuvre suggestive of an attempt to kill time – rather, one might say, she procrastinates herself round – until the north elevation faces south, when immediately she becomes a sausage again, turned about. All this is done with such perfect modesty that you immediately forget whether you saw her legs or not – indeed, whether she had any. As a matter of fact, I may here inform a doubtful public that Guy Fawkes has feet: her legs – if she has them – she, with propriety, veils in certain lashings of fat.

Bored observers watch a bored animal slowly heave itself round. But, as Morrison implies, thousands of people still came to see a hippo even after forty-four years of familiarity. The experience hadn't changed very much and always promised more than it delivered.

Another view of Guy Fawkes's immobility may be found in a piece about the zoo written in 1903:

> She is crippled by age and confinement. And spends most of her time lying motionless on the ground or else immersed in her bath with only just the end of her muzzle showing above the surface.[77]

While Bartlett saw Guy's desire to lie in her tank all day as sulky and Morrison saw it as lazy, here it is seen as the disablement brought on by years of an unnatural and constrained regime. And at this point she still had five years to live.

---

77 Elwes and Wood, 'The Zoo Past and Present'.

1 The Life and Times of Obaysch the Hippopotamus

The hippo that had been bought from Amsterdam came and went and in 1892 a male, called Jupiter, was purchased from Antwerp Zoo and arrived at the age of just under two on 5 July 1892.[78] By this time hippos were not the rarities they had been and my guess is that Guy Fawkes, the last remnant of Victorian hippomania, and now awkwardly living out her last days, like a dowager duchess becalmed in the new and racier Edwardian era in which she had decided to take no interest, ignored him.

---

78 In fact, Morrison gives the only account of Jupiter I have been able to find and characterises him as far from fully grown but already learning from Guy Fawkes the art of almost total inactivity.

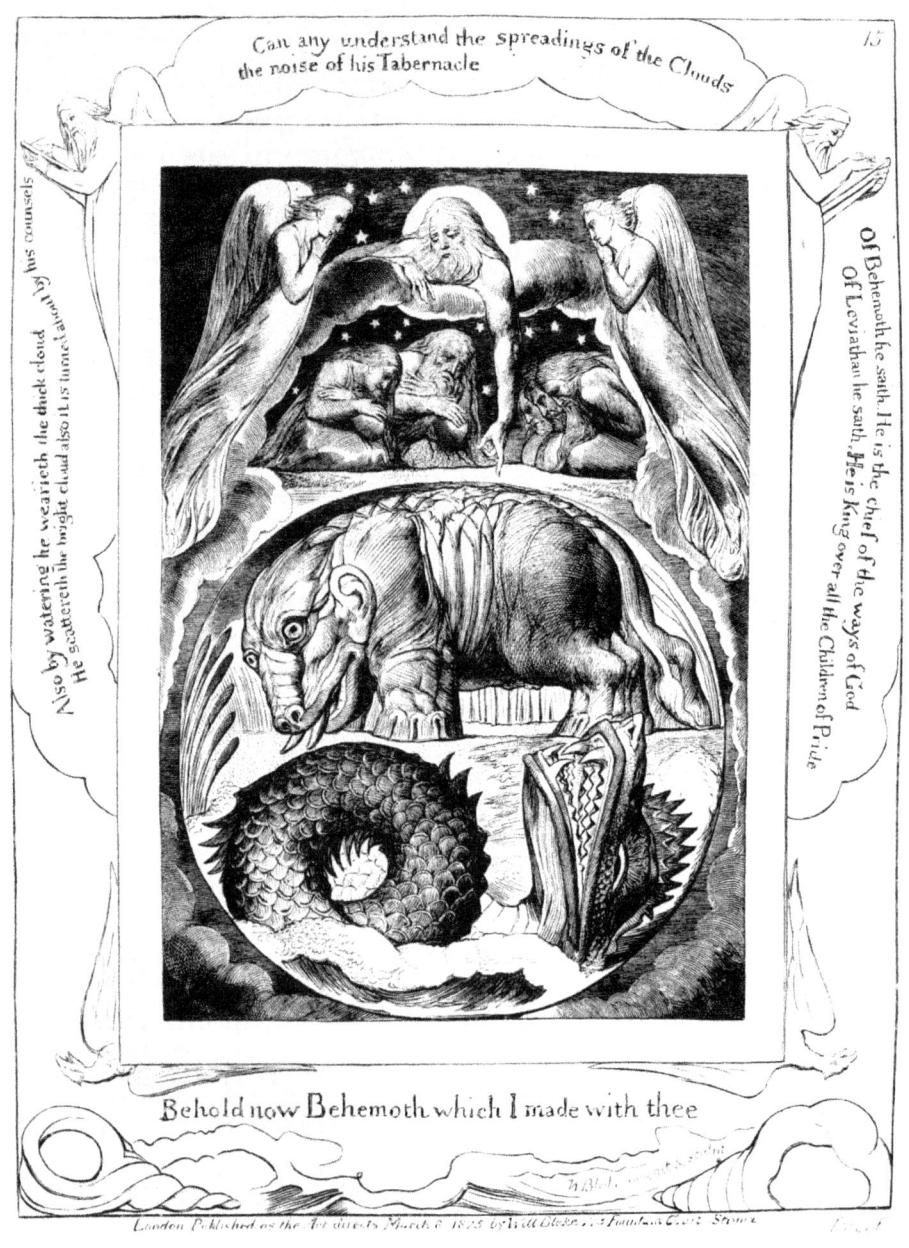

Plate 1 William Blake's Behemoth from his illustrations to the *Book of Job* (1826).

Plate 2 Late-eighteenth-century engraving of a hippopotamus. Although this kind of image was gradually replaced by the more accurate depictions of artists like Daniell, a version of it was still doing duty as late as 1875 (*Child's Companion*, 1 March 1875).

Plate 3 'Hippopotamus' engraving from the painting by Daniell published by Cadell and Davies (London, 1807). This, or subsequent versions of it, would have been one of the images that people coming to see Obaysch brought with them in their heads.

Plate 4 'The Haunt of Behemoth', *Illustrated Sporting and Dramatic News*, 21 November 1874. A late flowering of the tusked hippo image – by then most people knew that hippos didn't look like this.

Plate 5 Obaysch and Hamet. 'The Hippopotamus in the Gardens of the Zoological Society, Regent's Park', *Illustrated London News*, 1 June 1850. This is the earliest image to show Obaysch and his first keeper.

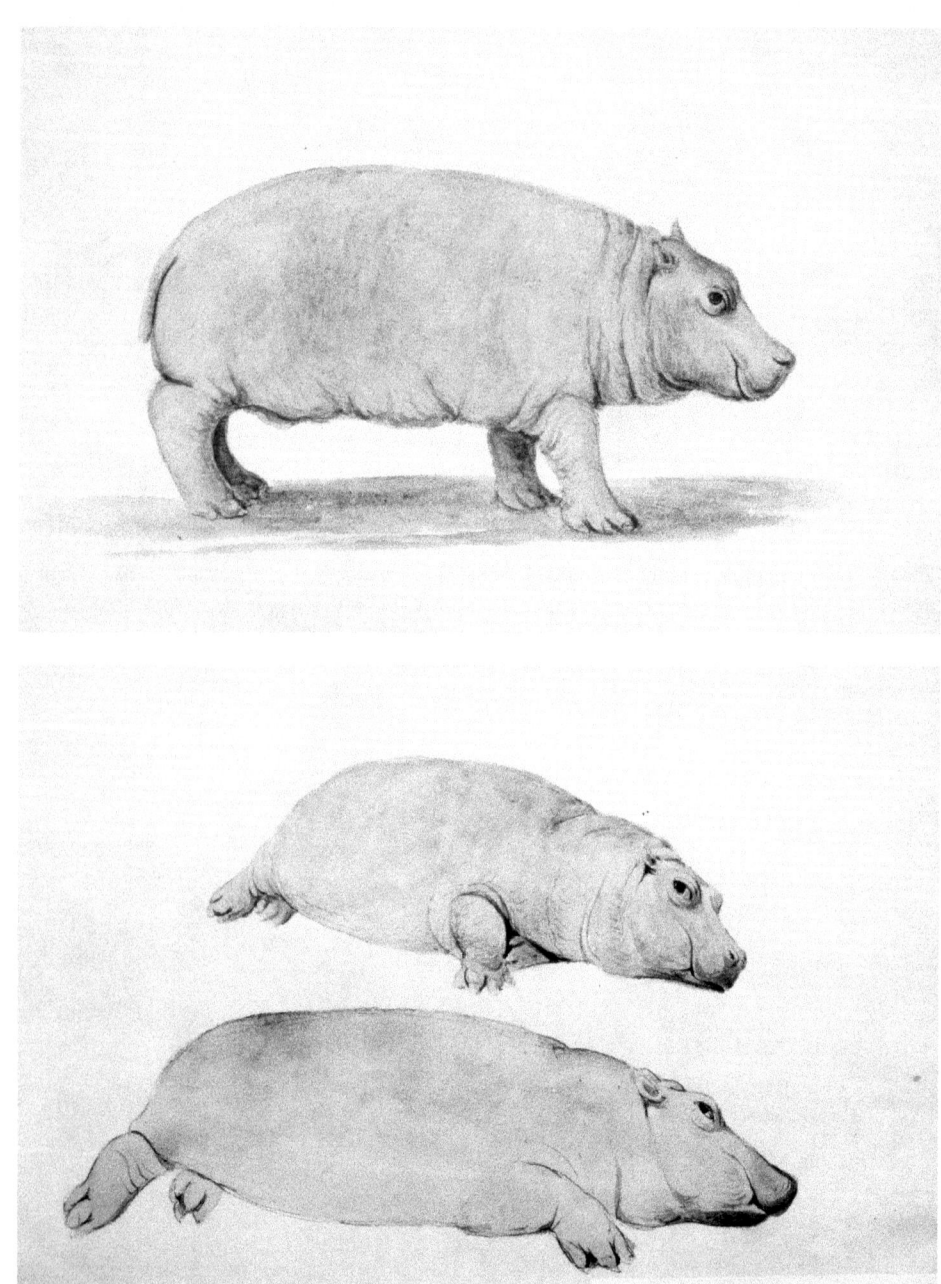

Plates 6 and 7 Watercolours of Obaysch by Joseph Wolf (1850), based on sketches sent from Egypt by Charles Murray. These show very clearly just how relatively small Obaysch was at the time of his capture. Wolf has also been assiduous in reproducing the complex tones of Obaysch's skin.

Plate 8 'The Hippopotamus in the Gardens of the Zoological Society, Regent's Park', *Illustrated London News*, 8 June 1850. This is by Joseph Wolf and clearly not drawn from life.

Plate 9 'The Hippopotamus in His New Bath in the Zoological Gardens, Regent's Park', *Illustrated London News*, 14 June 1851. An image of Obaysch on display, and of the experience of seeing him early in his captivity.

Plate 10 Obaysch meets Adhela: 'The Female Hippopotamus at the Zoological Gardens, Regent's Park', *Illustrated London News*, 12 August 1854.

Plate 11 'The Baby Hippopotamus at the Zoo', *The Graphic*, 1871. Here is one of the early attempts to construct an image of Adhela's maternal care for Guy Fawkes. Note the pillow – such a thing was also mentioned in the early accounts of Obaysch.

Plate 12 Guy Fawkes is now in motion but still tenderly overlooked by Adhela. 'The Hippopotamus and Her Young One at the Zoological Gardens', *The Graphic*, 1872.

Plate 13 'Sunday Afternoon at the Zoological Gardens – Beauty and the Beast', *The Graphic*, 21 November 1891. The hippo would have been Guy Fawkes and the lady is probably going to lose her hand or at least spoil her gloves.

Plate 14 'The Hippopotami in Their New Tank at Central Park', *Harper's Weekly*, 29 September 1888. An example of the display of hippos in another zoo.

Plate 15 'Mr Gordon Cumming's South Africa Entertainment – view of the Limpopo with a Herd of Hippopotami', *Illustrated London News*, 5 January 1856. Cumming's display of trophies from his hunting expeditions was immensely popular and supplemented by stirring images such as this, which contrasted greatly with the usually more sedate experience of seeing Obaysch and Adhela.

Plate 16 'Hippopotamus and Salee', *The Graphic*, 26 January 1884. This illustration accompanied a piece on the tribes of the Sudan.

Plate 17 'Hippopotamus Hunting in Angola, West Africa', *Illustrated London News*, 17 January 1880. As hippos became scarce on the Nile, hunters had to look further afield.

Plate 18 'Native Hunters Harpooning a Hippopotamus'. This image shows the kind of barbed spear and rope that would have been used to capture Obaysch. *Illustrated Sporting and Dramatic News*, 21 November 1874.

Plate 19 'Hippopotamus Shooting', *Illustrated Sporting and Dramatic News*, 21 November 1874.

Plate 20 'Native Hunters Hauling a Hippopotamus Ashore.' Note the attachment of the rope at just the point where Obaysch had his scar. *Illustrated Sporting and Dramatic News*, 21 November 1874.

Plate 21 'The Adventure with a Hippopotamus', source unknown. An archetypal image of David Livingstone keeping his head while all around are losing theirs.

Plate 22 'Canoe destroyed by a Hippopotamus on the River Zambesi, South Africa', *Illustrated London News*, 19 May 1866. The Zoological Society wanted to dissociate Obaysch from this kind of image of hippos.

Plate 23 'Boat Capsized by a Hippopotamus Robbed of Her Young', *Illustrated London News*, 7 November 1857. An interesting example of the attribution of some agency to an exotic animal.

Plate 24 'On the Victoria Nyanza: a Hippopotamus Attacks a Shooting Party', *Illustrated London News*, 29 August 1899. A stylised image that captures the enormous scale of the hippo.

Plate 25 A sketch by Robert Baden-Powell of himself shooting hippos, from his *Lessons from the Varsity of Life* (London: C. Arthur Pearson Ltd, 1933).

Plate 26 'The New Comer – a Sketch in the Depôt of an Importer of Animals', *Harper's Weekly*, 29 September 1888. Note the baby hippo peering from the crate. In fact this is a remarkably sparse collection. Most images of such dealers show animals packed to the ceiling.

Plate 27 'Young Hippopotamus (*liberiensis*) recently landed at Liverpool (now dead)', *The Graphic*, 29 March 1873. A very early image of a pygmy hippo and a reminder of the mortality rates of captured animals brought to Europe.

Plate 28 A late image of Guy Fawkes. 'The Home of the Hippopotamus', *Illustrated London News*, 11 February 1899.

2
The Several Meanings of Hippos

Obaysch was the first hippo that anyone had seen in England for a very long time but, of course, there were many English people who had seen hippos on their travels in Africa. And many more had seen pictures of hippos and read about them in a range of books and improving periodicals. So when Obaysch arrived, there was a well-established pre-existing discourse about hippos. Hippos already carried meanings and they signified a range of things. This chapter explores the way in which Obaysch was contextualised by preconceptions about hippos, and how he challenged them. It considers some of the things that people saw when they looked at him and how that physical and visual encounter was filtered – or indeed determined – by what they already thought they knew about hippos and the pictures of hippos or the stuffed specimens that they had seen or read about. The questions underlying this chapter are very simply: how did Obaysch infiltrate Victorian London; how did he become a mid-Victorian; and how and why, for a period intensively but to some extent for his whole life, he became a centre of attention.

Numerous newspaper and magazine articles describing Obaysch's arrival at the Zoological Gardens were quoted or alluded to in the preceding chapter. These gave detailed accounts, either serious or humorous, of Obaysch's journey, distilled the zoological science for the general reader, and stressed Obaysch's status as the 'first and only' hippopotamus, a niche that he was to occupy at least for the first three years of his captive life. The interest generated by the hippo was bound eventually to catch Dickens's eye as he surveyed the Victorian social scene in quest of topics for his journalism or ideas and images for his novels.

By 1850 Dickens had also caught the public's attention and was already becoming an arbiter of taste and value, to such a degree that he sometimes seems to us an infallible guide to the mentality of his era. Of course, no writer is truly an unproblematic guide, especially when they combine the two roles of journalist and novelist as Dickens did. But it is the case that Dickens had such a sure grasp of his readers' taste and the limits of their tolerance that his fiction and non-fiction offer a unique window into the relationship between author and audience. No other English writer presents such a carefully crafted appearance of unfiltered access to the fine textures of that place where things are just out of reach, and this effect can be so strong as to cause us to forget the scepticism that should inform all our critical reading and especially to forget Dickens' own interventions in the creation of his readers' taste.

I grew up knowing many people who were born as Victorians. Victorian coins were still circulating when I was a child and I bought sweets with blackened pennies minted in the 1880s. I have lived much of my life in houses built for Victorians: the wider doors to allow the passage of crinolines, the alcove on the landing to enable a coffin to be turned on its last journey downstairs, the deposits of coal dust beneath the floorboards in houses now without fireplaces. But what really happened before 1901 remains a tantalising mystery populated only by the Light Brigade and General Gordon. And, for me, by Obaysch.

2 The Several Meanings of Hippos

Who was reading about and visiting Obaysch? In 1850 Dickens had just founded a new venture, the magazine *Household Words*. This was, not surprisingly given his high reputation, an immediate success and by the end of the year had achieved an average circulation per issue of 34,500 copies. This compares with the average circulation of the *Times* during the same period, which is variously estimated at between 30,000 and 38,000 copies per issue. Of course, circulation figures always lag behind readership figures, as more than one person often reads each copy of the newspaper or magazine. If three or four people read each issue – a conservative assumption given the size of Victorian households, and consistent with Dickens's own conservative estimate in 1858[1] – so we can be reasonably confident in assuming that at least 100,000 people per week were reading *Household Words* by the second half of 1850. In 1850 the population of London was over 2.5 million and the population of the British Isles was about 27 million, meaning about 4 per cent of the population of London or about 0.4 per cent of the population of the British Isles were reading *Household Words*.

Compare this readership with the nearly 228,000 people who went to the Zoological Gardens to see Obaysch in his first few weeks: slightly more than 9 per cent of the population of London, and almost 1 per cent of the population of the entire nation. In the earliest days of his captivity in London, more people saw Obaysch than read about him. He was not yet the media construction he subsequently became, but an existential encounter.

When we speak of Obaysch as the centre of a craze we can, when we begin to look at roughly sketched figures like these, start to see what that actually means in terms of public participation, access and exposure. Yet the figures admit of contradictory interpretations. On the one hand, they suggest a significant interest backed up by thousands of individual decisions to have the new experience of seeing a hippo. On the other hand, they remind us that in Victorian

---

1 'The Unknown Public', *Household Words*, 21 August 1858, p. 217.

England the number of people who figured in cultural and consumer consumption was relatively small at the luxury end (i.e. the market for literary periodicals, serious newspapers, and hippos). In 1851 the most numerous classes of workers – the general labourer and the housemaid – earned an average of £44 a year and £12 a year respectively (in the case of the housemaid, various in-kind benefits enabled her to eke out her modest income considerably). If a general labourer wanted to see Obaysch, he would have to spend 0.1 per cent of his annual income; a housemaid with a yen to encounter a pachyderm would need to spend 0.4 per cent of hers. Compare that to a clergyman on an average annual income of £267, a doctor on £200 (an interesting example of how the relative value attributed to different professions has shifted), or a barrister on £1837 (no change there) and one begins to see the realities of the cost of seeing a hippo. *Household Words* cost just two pence per issue, but even that would have required some thought on the part of a housemaid with only four shillings and sixpence a week (fifty-four old pennies). She would have needed to invest nearly 4 per cent of her weekly income to buy a copy.

It is, of course, very hard to compare costs and income over time as purchasing power and economic value differ, and the relative availability of consumer goods changes the logic of individual purchasing decisions. It would appear, for example, that the Victorians spent far more on entertainment (music halls, theatres, etc.) than people did 150 years later, when the availability of mass-produced clothes, household appliances, recorded music, data, foreign holidays, etc. caused a shift in spending patterns attended by dark theatres and derelict zoos. But what the figures do suggest is that the choice to see or read about Obaysch involved a significant outlay for the vast majority of the population. When we speak of cults, crazes and manias, we are probably talking of the temporary enthusiasms of a relatively small section of the population, but one that was disproportionately important in cultural, economic and political life.

2 The Several Meanings of Hippos

In the very first volume of Dickens' new venture, the main subeditor, Richard Henry Horne, contributed a lengthy piece entitled 'The Hippopotamus'.[2] It appears that not only had Dickens noticed Obaysch, but he had also noted a rapidly rising tide of enthusiasm for hippos among the kind of people who constituted his core readership. This surely tells us a great deal about the place that Obaysch was already beginning to occupy in the Victorian imagination and the success of the Zoological Society's campaign to establish a profile for its new star animal.

Horne's piece is essentially a light-hearted and humorous account of Obaysch's capture, transportation and arrival in the Zoological Gardens. It includes some amusing made-up dialogue between, for example, the Pasha and his officers, and the Pasha and the consul-general. It is largely based on the official versions published in the *Times* and elsewhere but it includes one or two details not found in other accounts. These may be entirely fictionalised, or they may have been based on Horne's conversations with people at the Zoological Society who knew the inside story. The whole piece is really a prologue to its final paragraph:

> We are certainly a strange people – we English. Our indefatigable energies and matchless wealth often exhibit themselves in eccentric fancies. No wonder, foreigners – philosophers and all – are so much puzzled what to make of us. They point to the unaided efforts of a Waghorn, and to his widow's pension-mite – and then they point to our hippopotamus! Truly it is not easy to reply to the inference and impossible to evade it. We have had a Chaucer and a Milton, a Hobbes, and a Newton, a Watt and a Winsor, and we have had other great poets, and philosophers, and machinists, and men of learning and science, and have several of each now living among us: but any amount of a people's anxious interest, which the present state of popular education induces, is

---

2   R.H. Horne, 'The Hippopotamus', *Household Words* 1 (1850), pp. 445–449.

very limited indeed compared to that which is felt by all classes for a Tom Thumb, a Jim Crow, or our present Idol [i.e. Obaysch]. Howbeit, as the last is really a great improvement on the two former fascinating exotics, it is to be hoped that we shall, in the course of time, more habitually display some kind of discrimination in the objects of our devotion.[3]

Obaysch is here contextualised by the popular taste of which Horne himself, like Dickens, was simultaneously an architect, a builder, an arbiter and a critic. Obaysch is also seen against the backdrop of that growing model of Englishness that sees the English as overwhelmingly energetic, inevitably successful, but prone to (likeable) eccentricity and a tendency to prefer lowbrow entertainment over more challenging intellectual or cultural encounters. This is a self-image that has proved tenacious. Perhaps this is because it had, and still has, some truth. Perhaps it is because the model itself has shaped the behaviour of those it claims to represent. Either way, Obaysch is here put fairly and squarely within a pre-existing and readily comprehensible image of the English at play. By locating him in this way, Horne begins the process by which Obaysch himself would come to be seen as English and to be described in terms often used to explain English culture.

A few months later Dickens himself wrote a curious piece, entitled 'The "Good" Hippopotamus', for the second volume of *Household Words*.[4] This is a fantasy based on the premise that Obaysch's keeper, Hamet, has realised that his own fame is entirely dependent on Obaysch's and that, therefore, it would be appropriate to raise, by subscription, the money to build an equestrian statue of him – an idea that had already been mooted in an earlier issue of *Household Words*. The article is very far from being one of Dickens' most distinguished pieces of writing. It depends ultimately on the

---

3   Horne, 'The Hippopotamus'.
4   C. Dickens, 'The Good Hippopotamus', *Household Words* 2 (1850), pp. 49–51.

2 The Several Meanings of Hippos

rather flimsy joke that an Arab could organise a subscription. We saw earlier how Hamet was often dismissed as simply the creature of the hippopotamus craze, and this is another example. The piece doesn't add very much to our understanding of Obaysch beyond establishing his enormous popularity and beginning to set him up as a gentle giant:

> He was an easy, basking, jolly, slow, inoffensive, eating and drinking Hippopotamus ... he bathed in cool water when the weather was hot, he slept when he came out of the bath; and he bathed and slept, serenely, for the public gratification. People of all ages and conditions, rushed to see him bathe, and sleep, and feed; and H.R.H. [His Rolling Hulk] had no objection.

This also, perhaps, alludes to the growing sense that seeing Obaysch wasn't quite as exciting as all that, but it does portray him as inoffensive and placid and, as we have seen, there seems to have been a definite effort to ensure that episodes where Obaysch displayed his natural strength and aggression were kept away from the public eye wherever possible.

The impact of Dickens on the public memory of Obaysch can be seen some forty years later, when William Thornbury wrote that 'The hippopotamus which thus became a household word, for many years continued to be a prime favourite with the public.'[5]

This passage reminds us that although Obaysch gradually presented a lower media profile, his physical presence lost its attraction much more slowly. Obaysch also popularised a word ('hippopotamus') that would have been rarely used before, and it was Dickens who domesticated it through, ironically, *Household Words*. I am sure that Thornbury very deliberately used the phrase 'a household word' to refer his readers back to Dickens' foundational

---

5   W. Thornbury, *Old and New London*, 6 volumes, 5 (London: Cassell & Co., 1887–93), p. 258.

account. In fact, people were still going to see Obaysch in the 1870s and, by this time, a visit to see him had acquired the kind of formality appropriate to a celebrity. The various modifications to the hippo house, described in Chapter 1, had helped to manage and even to domesticate the experience. In 1875 Daniel Gorrie's comic travelogue *Geordie Purdie in London* – which recounts the amusingly blunt observations of a no-nonsense countryman from Fife – describes a visit to Obaysch (note that Gorrie assumes that anyone showing a visitor around London would take them to see the hippos), and his account of the physical context is especially interesting:

> [The hippo house has] quite a smart and elegant appearance imparted by the small paved court at the front door, the short flight of steps, the glass door, and the oil-cloth covered lobby leading into the interior where the huge animals are provided with all the requisites of amphibious existence.[6]

This is no longer a visit to a pair of hippos in a zoo, but a visit to a middle-class lady and gentleman entirely anthropomorphised by their quasi-domestic house with its approach, its doors, and its interior.

The depiction of Obaysch as a harmless and loveable lump sitting within a very specific bubble of Englishness that the two *Household Words* articles establish set the parameters for almost everything else that was said of him in public for the next quarter of a century. When people came to see Obaysch after 1851 (at least until the 1860s and perhaps beyond), many of them would have observed him through the lens given to them by Dickens.

---

6   D. Gorrie, *Geordie Purdie in London* (London: Houlston & Son, 1875), p. 109.

2 The Several Meanings of Hippos

When Obaysch arrived in Regent's Park, public knowledge of hippos was limited largely to scientific accounts, and to the accounts of explorers and hunters. It was generally, but by no means universally, thought that the hippo was the same animal as the Behemoth of the Old Testament, so Obaysch also had a place in the religious narrative. This was important, as a significant problem for Victorians as they made more and more discoveries of animals new to them was to decide why each animal had been created and what role it had in the bigger divine picture. When a Victorian looked at an animal she or he might well have asked – in a way which we would never do, and which is helpfully illustrative of the gulf between us when we start to fancy that the Victorians are little more than frock-coated or crinoline-wearing versions of ourselves – 'What is it for?'[7] In Obaysch's case the answer was, according to the book of Job: 'He is the chief of the ways of God'.[8]

The *Juvenile Missionary Herald*, which published a short article on hippos as one of the natural history notes with which it interspersed its other materials (on the grounds that young people set for the missionary life in Africa needed to know this stuff) entered into a dialogue with its readers on this text:

> Just read that passage of scripture in Job XL 15. Well, I suppose you have. What do you think of it? Is it not a wonderful description? It is generally thought that the 'Behemoth' is the animal of which our article gives the history.[9]

William Blake imagined Behemoth in his illustrations to the *Book of Job*. His Behemoth is a bizarre creature. Certainly one can see a hippo there, but there is also the carapace of a rhinoceros and the

---

7 George Perry asked this question in respect of the notably torpid koala (then called, among other things, the New Holland sloth) in his *Arcana; or the Museum of Natural History* (1811). See J. Simons, *Rossetti's Wombat*, p. 25.
8 Book of Job, 40 verses 15–24.
9 *Juvenile Missionary Herald* (1866), p. 28.

feet of an elephant. The animal has massive human ears and two tusks. Blake would never have seen a hippo, of course (unless it was the stuffed one in the Holophusikon) but he could have seen plenty of illustrations and many of these depicted the hippo's prominent and dangerous incisors as tusks.

The Portuguese explorer Father Jeronimo Lobo offered one of the earlier European descriptions of a hippo in his account of his travels, published in English in 1789 as *A Voyage to Abyssinia*. Lobo states categorically that the hippo has 'two tusks like those of a wild boar only larger.'[10]

Early engravings for the various editions and translations of Buffon's *Histoire Naturelle* (1749) frequently show these tusks, as does a much-reproduced illustration from Anders Sparrman's *A Voyage to the Cape of Good Hope, towards the Antarctic Polar Circle, and round the World: But Chiefly into the Country of the Hottentots and Caffres* (1785), which contained a much-cited description of hippopotamus behaviour.[11] A marvelous engraving in the *Penny Cyclopedia of the Society for the Diffusion of Useful Knowledge* (1838) shows a hippo with massive tusks happily munching away on a mouthful of grass – the text refers to these dental wonders as 'canines'. This illustration is based on the one by Landseer in John Barrow's *Characteristic Sketches of Animals Principally from the Zoological Gardens Regent's Park* (1832).[12] Of course, although most of the illustrations in Barrow's book were drawn from life, the hippopotamus could not be, and the text bemoans the fact that naturalists had hitherto been unable to benefit from a really good illustration of a hippo – which was now supplied. The vignette of a hippo's head that William John Burchell sketched from an animal

---

10   J. Lobo, *A Voyage to Abyssinia* (London: Elliot & Gay, 1789). This was originally translated from French by none other than Samuel Johnson.
11   A. Sparrman, *A Voyage to the Cape of Good Hope* (London: G.G.J. & J. Robinson, 1785).
12   J. Barrow, *Characteristic Sketches of Animals Principally from the Zoological Gardens Regent's Park* (London: Moon, Boys & Glaves, 1832).

2 The Several Meanings of Hippos

killed by a group of bushmen, which he published slightly before Barrow's book appeared, in his *Travels in the Interior of Southern Africa* (1822–24) offers a good likeness, without tusks.[13] But the image in Sir Andrew Smith's *Illustrations of the Zoology of Southern Africa* (published in parts between 1838 and 1849) certainly supplied the want, and appeared just before visitors to Regent's Park could see a hippo with their own eyes.[14] It is a delicate portrayal showing a fully grown hippo (without tusks) and her infant, who is looking to suckle. Some of the images of Adhela with Guy Fawkes seem to owe something to this picture, as do the watercolours of the young Obaysch that the Zoological Society commissioned from Joseph Wolf (who went on to become their *animalier* of choice). Wolf's painting were based not on life drawings but on sketches provided by Murray in Cairo when Obaysch was living in his garden. Murray was certainly the sort of man – he had both the inclination and the money – who would have been a subscriber to Smith's *Illustrations*. It seems to me, therefore, that Wolf's painting has a debt to Smith either through Wolf's own knowledge of the image or through the influence Smith's work may have had on Murray's sketches. This shows especially in the somewhat sentimental portrayal of the animal. This is a new strain in hippo imagery and contrasts with the tusked savagery that had hitherto dominated representations of the hippopotamus.

The idea that a hippo's teeth were like tusks was important to the construction of the image of the hippopotamus just prior to the arrival of Obaysch. Once Obaysch was in London, everyone could see that the teeth were not tusks and did not protrude to anything like the extent that some illustrations had led people to believe. The tusked hippopotamus was imagined as a fierce and dangerous animal and as the nineteenth century progressed, when

---

13   W.J. Burchell, *Travels in the Interior of Southern Africa* (London: Longman, Hurst, Rees, Orme & Brown, 1822).

14   A. Smith, *Illustrations of the Zoology of Southern Africa* (London: Smith, Elder & Co., 1838–49)

people wanted to illustrate the dangers of unexplored Africa they frequently turned to the hippo, lurking just below the surface of the river, waiting to capsize a boat and snap the canoeists in two, rather than to the lion or the crocodile. For a maritime nation like the British, a creature that specialised in sinking vessels would have been a particularly terrifying signifier of disorder. In Africa the boat or canoe was a place of safety. Natives could not ambush you so easily when you were on the river and the water offered a more rapid and secure mode of transport than going on foot or being carried by animals or African bearers. So the hippo represented an ultimately disruptive force to the progress of discovery and imperial expansion, and so to the spread of civilisation. Wild hippos were thus configured as both a physical danger and an ideological obstacle. They were not only uncivilised; they were also a threat to civilisation, which makes the construction of Obaysch within the paradigms of Englishness and domesticity all the more interesting.

Early illustrations of the hippo may have wanted in accuracy, but there were numerous scientific accounts available to the reading public and to the gentlemen of the Zoological Society. Collectively, these narratives constituted a reasonably accurate picture of some of the habits and behaviour of hippos in the wild. Dampier is an odd man out here, as it appears he confused the South American tapir with the hippo, but that is perhaps understandable. As we have seen, the confusion (perhaps solely lexical) continued well into the nineteenth century.[15] Buffon erroneously believed that hippos were

---

15 The Malayan tapir (as opposed to the New World tapir) was first described in 1819 by the French naturalist Desmarest. However, William Farquar had sent a full description to the Asiatic Society in 1816. Unfortunately for him they didn't get round to publishing it until 1821. In the meantime, Sir Stamford Raffles (in a rather unseemly manner) tried to get his own description published and always discounted Farquar's claim to be the first English

unable to travel very fast on land (intuitively this sounds fair enough but, as we saw in the previous chapter, at least one of Obaysch's keepers had occasion to test empirically the falsehood of this claim when he was pursued by an angry hippo at full gallop). By and large, however, there was a zoological literature from which Obaysch's captors and keepers could draw some hints on how to maintain their rare and valuable star. They did not know about the hippo's lifespan and breeding cycle, and did not know enough about nutrition, but they were not working in complete ignorance. Interestingly, the belief that hippos might not be able to survive if they had only sea water to bathe in, which led to the massive effort to maintain a fresh water pool on the S.S. *Ripon*, shows that Obaysch's keepers were not prepared to trust Sparrman's account when it came to the care of their hippopotamus. Sparrman had described beach-living, sea-going hippos. Maybe Obaysch's captors thought that as he had been captured in fresh water he would need fresh water to survive. Similarly the account of a hippo hunt by Henry Salt (one of 'Hippopotamus' Murray's predecessors as British consul-general in Egypt) describes fairly accurately the rising and falling of hippos as they slept and breathed in a river. Perhaps this is why Bartlett was so fearful when Guy Fawkes appeared to have been under for too long.[16]

Carl Peter Thunberg's *Travels in Europe, Africa and Asia Made between the Years 1770 and 1779* (published in English in 1793) contained an account of the attempted capture of a hippo that was

---

naturalist to describe the tapir. When Raffles was sailing for England in 1824 a fire on the ship led to the loss of his entire and extensive natural history collection, including many live animals (including a tapir). It was this traumatic event that led him to propose the foundation of the Zoological Gardens. See V. Glendinning, *Raffles and the Golden Opportunity* (London: Profile Books, 2012), H.J. Noltie, *Raffles' Ark Redrawn* (London: The British Library, 2009) and J. Bastin and K.C. Guan, *Natural History Drawings* (Singapore: National Museum of Singapore, 2010).

16 J.J. Halls, *Life and Correspondence of Henry Salt* (London: Richard Bentley, 1834), pp. 207–210.

quoted in William Bingley's popular *Animal Biography, or, Authentic Anecdotes of the Lives, Manners and Economy of the Animal Creation* (1803):

> Professor Thunberg was informed, by a respectable person at the Cape, that as he and a party were on a hunting expedition, they observed a female hippopotamus come from one of the rivers, and retire to a little distance from its bank, in order to calve. They lay still in the bushes till the Calf and its mother made their appearance, when one of them fired, and shot the latter dead on the spot. The Hottentots, who imagined that after this they could seize the Calf alive, immediately ran from their hiding-place, but though only just bought into the world, it got out of their hands, and made the best of its way to the river, where, plunging in, it made safely off. This is a singular instance of pure instinct, for, the professor observes, the young creature, unhesitatingly, ran to the river, as its proper place of security, without having previously received any instructions from the action of its parent.

I have questioned the veracity of the account generally given of Obaysch's capture and offered an alternative version. Here, in Thunberg, we find a possible source for that original account. This source enabled the London zoologists to place their new acquisition squarely within a pre-existing and authoritative European tradition of writing about hippos. They could thus begin, from their very first encounter with Obaysch, a process of familiarisation by which the exotic was placed within a tradition contained by well-worn scientific knowledge on the one hand, and the established tropes of African travel and discovery on the other.

The Renaissance naturalist Pierre Belon wrote of seeing a hippopotamus in his *L'Histoire Naturelle des Estranges Poissons* (1551). This animal (and there has been some doubt as to whether it really was a hippopotamus, although Belon's account does make it sound very much like one and it isn't easy to imagine what else it

2 The Several Meanings of Hippos

may have been) was kept with the Sultan's elephants in the ruins of the palace of Constantine:

> When some foreigner comes to see the aforementioned Hippopotamus, they show it to him, if he gives them some money. They make it come out of its stable unbound, having no fear that it might bite. Then its keepers, wishing to please whoever they are showing it to more, have some head of round cabbage or melon, or a handful of grass, or, indeed, some bread, given to them, which they hold up in the air, showing it to the Hippopotamus. Understanding that they want it to open its mouth, it opens it so widely that the head of a yawning lion could fit inside it. Then its keeper throws it what he has shown it, as if he were throwing into a large sack. The Hippopotamus chews this and then swallows it.[17]

This account of how to show a hippo in order to maximise income is not so very much unlike what was happening to Obaysch. What is especially striking is the idea of the docile hippo trotting after its keepers and being thrown treats. This is something we have seen in the accounts of how Hamet worked with Obaysch and, as we have seen, these accounts may well have been somewhat coloured in order to play down Obaysch's natural aggression and the coercion that appears to have been a part of his captivity in the early days. But here is a literary source from one of the founding fathers of modern ichthyology, and this further enables that process of familiarisation by which Obaysch was settled into the semantic economy of the zoo as well as into its *lexis* and made meaningful to those who saw him and read about him.

---

17 Pierre Belon quoted in J.B. Lloyd, *African Animals in Renaissance Art* (Oxford: The Clarendon Press, 1971), pp. 78–79. Belon produced the first image of a hippo to be published in France and also a wonderful illustration of the Roman Emperor Hadrian engaging in the risky business of swimming in the Tiber with his pet hippo Antonius. See R. Cooper, *Roman Antiquities in Renaissance France* (Farnham, UK: Ashgate Publishing Ltd., 2013), p. 161.

What we see in the imagining of hippos, even after Obaysch could be seen in the flesh, is an awkward balancing act. There was a significant but partial body of reasonably accurate zoological information. There was, at the same time, much to be learned about hippos, but the scientists of the Zoological Society knew all too well (these were men of great ability and intelligence after all, and we should never underestimate either their scientific acumen or their desire to maintain their animals in the best possible conditions) that they could not learn very much from a hippo who lived in a small pond in the middle of London. In some ways the ability to perform autopsies occasioned by the deaths of Obaysch and Adhela's first two children enabled them to learn more than they could from observing their captive hippos. In addition, although the notion of Orientalism had not yet been constructed as a fully articulated theoretical concept, it already flourished as an aesthetic discourse, both visual and literary, with a range of readily deployable terms and tropes within which to place new objects or to interpret new phenomena.

Africa, however, was different. The Roman naturalist Pliny the Elder's notion that out of Africa there is always something new (*ex Africa semper aliquid novi*) still seemed to hold true for the Victorians.[18] There were two reasons for this. The first was that large areas of Africa had yet to be fully explored or even visited by Europeans and were still more or less blank on European maps. The second reason, which partly derived from the first, was that there was no discourse of 'Africanism' to enable the ready organisation of African things into a pictorial and aesthetic grammar. Things Turkish, Arabic, Central Asian or Indian were assimilated via the well-worn paths of the Orientalists and Orientalism, but there was no equivalent category of 'African'. This is why, I suspect, that so

---

18   The actual quotation (from Pliny's *Historia Naturalis*, VIII, xvii) is '*semper aliquid novi Africam adferre*'. For the Victorians as for the Romans, understanding the natural history of Africa was a work in progress and still capable of throwing up the most unexpected things.

2 The Several Meanings of Hippos

much was made of Obaysch's Egyptian origins and the Arabic or Turkish appearance of Hamet, who, as I have suggested previously, was clearly, like Obaysch, an African – and in fact the 1852 edition of the Zoological Society's *Description of the Gardens and Menagerie* explicitly refers to him as 'a negro'. Casting Obaysch as Egyptian enabled him to be understood within an existing framework of ideas, and so to be more readily assimilated into a framework of Englishness and tamed, as the Orient had been tamed, at least discursively and aesthetically, by a set of easily understood images that had themselves been incorporated into the mental world of English culture.

Africa did not offer the same opportunity. Insofar as one can speak of a discourse of Africanism at this time one can only speak of it in terms of the unknown, the dark, the undiscovered, the primitive, the violent, the cruel, and the savage. The cunning but sophisticated and noble Arab is set against the benighted and childlike African. Bishop Reginald Heber's once famous hymn 'From Greenland's Icy Mountains' (1819) referred to 'Africa's bright fountains' but also, in its original draft, to 'the savage in his blindness' who 'bows down to wood and stone.' I have little doubt that Heber had in mind the African, in particular, when he was envisaging the challenges and triumphs of missionary work. The picturesque slavery of the Ottomans, with all the opportunities it afforded for prurient enjoyment and the development of the pornographic imagination, is set against the abject misery caused by the sub-Saharan trade. The second Lady Baker, wife of the soldier, explorer and big game hunter Sir Samuel Baker (who bagged his fair share of hippos over the years), was first encountered by him in a slave auction in the Bulgarian town of Vidin. One cannot imagine any Victorian traveller, even one as adventurous and unconventional as Sir Richard Burton, buying a black slave in an auction in, say, Zanzibar, and then marrying her.[19]

African animals were largely dangerous and were still being discovered. Europeans appeared to die in every way possible: most often from malaria and yellow fever but also from diseases that nobody understood or could even name or from bites from insects that had yet to become part of the taxonomic systems of western entomology. F. Harrison Rankin's *The White Man's Grave – A Visit to Sierra Leone in 1834* (published in London in 1836) with its glum epigraph, 'It is quite common to ask in the morning, how many died last night', sums up the dread that Africa, especially west and central Africa, could inspire. The well-organised impis of the Xhosa, Matabele and Zulu in the south were easy to understand as they charged across the veldt towards you in impeccable formation but the Ashanti, seen only through the dim and transient gleam of a golden ornament as they lurked in the rain forest, or the fearsome female soldiers of Dahomey, offered challenges which were not easy to contain within established frames of reference.

On the religious and cultural fronts the coherent and comprehensible monotheism of Islam was opposed to a fragmented and often fearsome animism – Bishop Heber's wood and stone. Even the alleged predilection for sodomy – 'pederasty', as our recent ancestors were pleased to call it – among the Turks and Arabs within Sir Richard Burton's 'Sotadic Zone' gave once again an opportunity for prurient acknowledgement of a different but nevertheless recognisable set of cultural norms, even if those gave rise to horrible practices.[20] But this was set against the alleged threats of violent anal rape of Christian converts by a central African King.[21]

---

19   The most recent biography of Samuel Baker is M.J. Trow's *The Adventures of Sir Samuel White Baker, Victorian Hero* (London: Pen & Sword Books, 2010). P. Shipman's *To the Heart of the Nile* (New York: Morrow, 2004) is a fascinating biographical account of Lady Florence Baker. See also T. Jeal, *Explorers of the Nile* (London: Faber & Faber, 2011).
20   The Sotadic Zone was a region where 'pederasty' was common and, indeed, celebrated. Burton sketched this out in the 'Terminal Essay' attached to his 1885 translation of the *1001 Nights*. The zone encompasses both shores of the Mediterranean, much of Asia and the whole of the Americas, but not southern Africa or Australia.

So Obaysch floats between four distinct discursive systems. There is a world of scientific discovery and zoological exploration and research (his European dimension). There is a world of sentiment in which he becomes a 'sight of London', a husband and a querulous but much loved subject of Queen Victoria (his English dimension, as initially articulated by Dickens and *Punch*). There is a world of the exotic visitors from Araby, snake charmers, Hamet, old 'Hippopotamus' Murray chatting away to him in Arabic (his Orientalist dimension). Finally there is a world of violence, of teeth broken in mindless rages against iron bars, headlong pursuits of keepers and brawling with his mate (his African dimension). Everything we see and read in the depiction of Obaysch is essentially a contest for his meaning played out between these four competing discourses. There is no doubt that when we look at the picture overall we see that of all these discourses it is Obaysch's African dimension that is most often subordinated to the demands of the others and which is the least well accommodated to the nineteenth century's requirements for the representation of this curious and complex animal. Above all things it was crucial that Obaysch must never appear to be a savage. There was a great deal of money at stake but, more subtly, the accommodation of Obaysch within discursive structures that underplayed or concealed his African origins. Where they became visible, taming them became part of the growing business of Africa's colonisation.

In her excellent study of celebrity animals in the Victorian Zoological Gardens, Narisara Murray deploys the notion that these animals were the inhabitants of borders.[22] There is much to concur with in her argument and we can see how Obaysch may have stood on, or been contained by, at least four borders. I'd also add to Murray's insight the thought that such animals (perhaps all captive

---

21 On this episode see D.A. Low, *Buganda in Modern History* (Berkeley: University of California Press, 1971), pp. 27–29.
22 Murray, *Lives of the Zoo*.

animals on display in zoos, both in the nineteenth century and now) are also liminal – perhaps, more correctly, liminoid. They stand on a threshold between their own 'natural' status as a wild creature from an exotic place and their social and aesthetic construction as an object to be looked at and understood via more familiar cultures and the various disciplines of natural history – from sketching in water colour to ecology. These things make them a commodity to be consumed either as a spectacle or through secondary objects and representations. When it enters the zoo each animal crosses a threshold, which, it appears, can be of greater or lesser complexity depending on the rarity of the animal, the pre-existing discourses that structure it for its audience, and its impact on the viewer (compare, for example, the reactions of zoo visitors to a kangaroo with their reactions to a silver back gorilla: curiosity and amusement in the first case gives way to spontaneous gasps of awe in the second).[23] This impact can be intense and an animal like Obaysch, who was at the high end of the scale of both rarity and impact, clearly promised maximal intensity even though the experience of seeing him in the flesh (especially when he was young) could be disappointingly tame.

Although endeavours were made to place a *cordon sanitaire* around Obaysch's African origins, no such effort was made for Adhela. We have already seen that there was a greater preparedness to acknowledge her natural aggression than there was to acknowledge Obaysch's. Similarly, there appears to have been no reticence in casting her as an African. A *Punch* cartoon commemorating the birth of Guy Fawkes shows Adhela being visited by the queen lioness and her cubs. The lioness is dressed in European clothes and has an ermine stole and a crown on her

---

[23] Perhaps the best known study of this phenomenon is B. Mullan and G. Marvin, *Zoo Culture* (Urbana: University of Chicago Press, 2nd ed., 1999), See also D. Hancocks, *A Different Nature* (Berkeley: University of California Press, 2001).

2 The Several Meanings of Hippos

head. Adhela is dressed in the clothes that became the 'Mammy' stereotype, complete with spotted headscarf. The caption reads:

> Queen Lioness: And how is the darling, my dear Madame Hippo?
> Madame Hippo: Oh him berry well, tank you ma'am. Spects all de worl' comin' to see dis Baby![24]

What we see here, when we compare this and the issue of aggression to the treatment of Obaysch, is a grafting of a gendered perspective onto the delicate question of Obaysch's racial origins. There is no question but that Adhela is an African (albeit portrayed using an African-American stereotype at a time when most Americans of colour were still slaves, it might be noted, and thus perhaps lifting a veil on the zoo's subliminal understanding of the status of its animals). It has to be concluded that it was somehow less damaging to the balance of discourses described above to feminise negritude than to allow the notion that the male hippo was, somehow, black.[25] Indeed, stereotypes of the African male were, in some ways, challenged by Obaysch's actual behaviour. He clearly was not sexually predatory. He was certainly violent but this was played down or concealed until after his death. The single enduring trait that Obaysch consistently shares with Victorian stereotypes of the African man is his indolence, and this is easily swept up in a good-natured portrayal of the hippo as a gentle giant without any requirement to locate it within the terms of a commonplace racial and racist discourse.

---

24 'The Two Mothers: A Visit of Sympathy', *Punch's Almanack for 1873*.
25 The question of African-Americans and their proxy depiction via zoo animals also comes up in early treatments of the next 'star' animal, the American ant-eater. *Punch* (22 October 1853) shows him dressed in an early version of what became the familiar 'Uncle Sam' costume with a whip-like tongue flailing at a crowd of miniature African slaves.

So what did Victorians expect to see when they first encountered a hippo and, specifically, Obaysch? I have already mentioned that they might well have expected to see a creature with tusks. Being familiar with the Biblical Behemoth they would most likely have expected to see something very big, even though they knew Obaysch was still little more than a baby. They might well have expected something very fierce, as they would have been used to seeing images of hippos attacking canoes by overturning them or even biting right through them in popular biographies of Livingstone and other such sources. They may have seen images of hippo-hunting, including the use of barbed harpoons by natives and guns by Europeans. They might have expected to see something lively, as they would not have understood the predominantly nocturnal nature of the animal. As I have pointed out in the previous chapter, it appears that Hamet did his best to fulfill at least that expectation by prodding Obaysch into more activity than he was ready for. Obaysch featured in a typically large-scale advertisement for the celebrated clothiers and tailors Moses and Sons, who had the innovative advertising technique of associating their products with famous events of the day in verse. Their choice of Obaysch which shows how famous he was but obviously nobody from the firm actually went to see him and their poem fell at the first when it referred to his 'trunk.'[26]

When we collect the images and written descriptions that the best informed visitors to the Zoological Gardens would have read or seen we can start to see that Obaysch didn't fit many of them very well. The passage of time would solve the problem of size, of course and his aggression was not only concealed but also contained by the circumstances of captivity. In some ways, the taming of a big, fierce animal represented an important stage in the asymmetrical power relationship between colonial expansion, exploration for acquisitive

---

26 By 1851 this company, which had developed the concept of ready-made clothes, owned the biggest shop in London and were noted for their extensive and innovative advertising. It is another indication of Obaysch's prominence that they should have used him in one of their campaigns.

2 The Several Meanings of Hippos

purposes, science and Western thought and the objects of these processes. They are reified all too obviously in the body of a little hippo being gawped at by thousands of people in the middle of London. Clearly, it was necessary for the zoo to manage the aggressive traits of its animal inmates simply as a safety measure, but the process of taming has a less straightforward implication when it is contextualised by the role played by the often mutually articulated processes of hunting, science and exploration in the broader imperial project, especially as it was beginning to unfold in sub-Saharan Africa. The question of the hippo's nocturnal habit and its consequent lack of day-time activity appears to have been solved by imposing an unnatural regime.[27] Almost all the descriptions we have of Obaysch and the other hippos in the zoo – the narratives of the births of Adhela's babies are a notable exception – take place during the day, and one of the few night-time stories (the one in which the keeper went for a post-pub swim, only to find Obaysch resting at the bottom of the pool) emphasises the topsy-turvy world into which Obaysch was thrown. Hippos would not normally be in the water at night; they would be out foraging on land. In fact, I wonder if Obaysch's apparent poor health may have been caused not so much by the long-term impacts of the trauma of his capture, or by barely adequate nutrition, the lack of roaming room or dental disease, so much as by the enforced reversal of a hippo's natural biorhythms.

For those who had seen Obaysch in the flesh, describing him posed challenges. The first was how to relay his size: although people

---

27 The question of taming animals for display in zoos was an important one. The menageries, which did not have permanent buildings when they were touring, always ran the risk of an escaped animal, so forms of training were often employed beyond those used for the animals in the various circus-like acts the menageries also offered. See F.C. Bostock, *The Training of Wild Animals* (New York: Century Co., 1903) to learn the methods of a well-known Victorian lion tamer and menagerie keeper. D.A.H. Wilson, *The Welfare of Performing Animals: A Historical Perspective* (Heidelberg: Springer, 2015) offers a comprehensive scholarly account.

knew that adult hippos were very large, they were not so sure about a baby hippo. The two most commonly deployed comparisons were with a large dog, such as a Newfoundland, and with a pig. David Livingstone himself referred to seeing baby hippos in the wild 'not much larger than spaniels', so this comparator had an impeccable provenance.[28] The dog comparison had to do with size only, of course, while the pig comparison reflected an attempt to give some idea of Obaysch's shape as well. Henrietta Halliwell-Philips went to see Obaysch on 22 June 1850 and noted in her diary that he was 'a short thick heavy animal something like a pig about the mouth & of a dark brownish colour.'[29]

Obaysch's colour was harder to pin down. His body was dark, Indian ink in tone, but there were paler, pinker areas, especially around his eyes, and occasionally the effusion of the garnet-coloured bloodsweat gave him a shiny red tone. His texture was not like leather but like India rubber. Indeed, he was frequently seen as an animated rubber ball. Of course, professional scientists like Richard Owen offered thorough descriptions of Obaysch's external morphology but, for the general reader, to imagine a large piebald dog or a pig made of gently shining rubber was sufficient to conjure up this exotic sight. And, of course, the very banality of the comparisons helped to pull Obaysch into the realm of the familiar and to begin the all-important process of domestication and Anglicisation that would establish him as a favourite sight. The managers of the Zoological Gardens had a difficult balance to strike. On the one hand they had to offer a 'star' animal that was mysterious and rare and representative of the cutting edge of their

---

28　See, for example the use of Livingstone as an authority in 'The Hippopotamus' (*The Child's Companion*, 1 May 1875). It is hard now perhaps, to understand the authority that Livingstone and his image would have commanded to the general public at this time, given the debunking that Livingstone received from Lytton Strachey and the fact that even sympathetic modern accounts of his life, like Tim Jeal's, have to admit that for all his years of dogged effort in missionary activity he probably never converted anyone.

29　Quoted in Ito, *London Zoo and the Victorians*, p. 121.

trade as zoologists. But, increasingly explicitly, they were also the proprietors of an entertainment venue. This meant that, on the other hand, they needed to make these animals familiar, even loveable, so that people would want to see them over and over again. Their genuine commitment to a scientific and educational mission from time to time conflicted with the business imperatives of tourism. In the fault lines between the scientific descriptions of Obaysch, the more discursive or journalistic descriptions, and the actual process of his capture and display we can perhaps observe the working-out of this dialectic.

A typical description of Obaysch in his mature years may be found in an article by Elizabeth Lawrence entitled 'A Garden Party of Wild Animals', written for children in 1874. She gives readers a guided tour of the Zoological Gardens with illustrated commentary on a selection of the animals. Here is what she saw when she got to the hippos:

> He [the rhinoceros] and the hippopotami are provided with huge baths of water inside their houses, for winter use, as well as the tanks in their yards, and the hippopotami spend as much time in the water as out of it. Beside the two old ones, there is a baby hippopotamus called Guy Fawkes because he [sic] was born on the fifth of November. Little Guy, being smaller and more active, is not quite so ugly as his huge father and mother, though all are hideous enough, with their pig-like bodies and horrid faces, with mouths that stretch literally from ear to ear, showing, when open, the whole roof of the mouth, the top of the head seeming to fold back like the lid of a box on hinges. These creatures are as vicious as they are ugly, and apparently entirely incapable of affection or intelligence.[30]

---

30 Elizabeth Lawrence, 'A Garden Party of Wild Animals,' *St Nicholas Scribner's Illustrated Magazine for Girls and Boys*, 1 August 1874.

Oddly the illustration accompanying this passage – which is of a wild bloat and not the zoo animals – is entitled 'The Hippopotamus and Her Baby' and shows an adult hippo in the river with a baby balanced on her back in what one might say is a rather affectionate image. But Lawrence stresses the ugliness, which had been a feature of writing about Obaysch since he arrived in London. The idea that hippos were like pigs was continued, and Lawrence obviously had some inkling that hippos were not cuddly old grumps but seriously dangerous animals.

This account contrasts sharply with another text for children written at much the same time and also accompanied by a picture of Adhela and Guy Fawkes, this time showing them in their enclosure. The written commentary stressed the baby's vulnerability and the need for nurturing:

> Here you see Mrs Hippopotamus and her baby when it was only eight months old; but now it has grown so very much that one can scarcely think it ever was a little helpless creature, needing the greatest care in order to preserve its life.[31]

This is a good example of how interest in hippos could be exploited in order to promote various causes. It shows how the bodies of hippos could be the repository of virtually any meaning required to make a point. The text was published in a magazine catchily entitled *The Children's Treasury and Advocate of the Homeless and Destitute*, and the writer was clearly using Adhela and Guy Fawkes not only to inculcate a set of caring values into juvenile readers but also to frame Guy Fawkes alongside other youngsters who need and deserve care and nurturing if they are to survive.[32]

---

31 'Mamma and Baby', *The Children's Treasury and Advocate of the Homeless and Destitute*, 13 March 1872.
32 B.T. Gates, in *Kindred Nature* (Chicago: Chicago University Press, 1998), offers an interesting account of the gendering of interest in natural history during the Victorian and Edwardian eras and, in particular, shows how the

2 The Several Meanings of Hippos

Hippos are also very noisy animals, although given that their vocalisations are largely designed to communicate with other hippos it may be that Obaysch was a relatively silent baby, which would account for the relatively few references to what he sounded like. In the early days we read about grunts when he was in Hamet's company but not much else. *Punch* spoke of 'a melancholy whine' whenever Hamet was absent and of Obaysch's 'tendency to blubber'. Given the more respectable commentary on Obaysch's apparent attachment to Hamet we might sadly acknowledge that one of the few noises he did make when he was first caged was a distress call. By the time Adhela arrived there was more chance to hear what a hippo sounded like and this was not something that impressed the listener:

> Some of our readers have seen the Hippopotamus at the Zoological Gardens, and, doubtless, have heard its discordant snort.[33]

Adelha is reported hissing 'loudly like a big snake' whenever she thought anyone was getting too close to the newborn Guy Fawkes. The past is always easier to sum up as a patchwork of visual images, and less so as a tapestry of sounds, smells, tastes and textures.[34]

---

idea of nurturing became a key concept in constructs not only of nature itself but also, more importantly, in the ways humans should respond to it. This concept also drives a rationale for science which is quite different from that which we might find in the approach of, say, Richard Owen (chosen here simply as a prominent representative of the Victorian male scientific community). See also M. Ferguson, *Animal Advocacy and Englishwomen 1780–1900* (Ann Arbor: University of Michigan Press, 1998).

33 *Chatterbox*, 27 July 1870.
34 One attempt to do this is J.M. Picker, *Victorian Soundscapes* (Oxford: Oxford University Press, 2003). See also A. Corbin, *Village Bells: The Culture of the Senses in the Nineteenth-Century French Countryside* (New York: Columbia University Press, 1998) and B.R. Smith, *The Acoustic World of Early Modern England* (Chicago: University of Chicago Press, 1999).

Cartoons of Obaysch from 1850 depict him in a number of ways but always as gentle and genteel. Four from *Punch* stand out in particular.[35] One shows a well-dressed young woman leading a hippo on a ribbon and saying, 'Come along, Fido.' The hippo's tail is decorated with ribbons and he has a little bell around his neck. He is about as tall as his companion's waist and has a smile on his face. The second, 'The Sea-side Season', shows a bathing machine being wheeled down into the sea from the beach and a hippo wearing a massive bonnet and delicate sandals climbing up the ramp. This is supposed to represent Obaysch taking a much-needed holiday:

> The fashionable lions will soon be 'running down' to the sea-side, and if such refreshment is required for the 'fashionable lions,' why not for that greatest of all lions of the season, the Hippopotamus? We think it is high time that the poor animal obtained the benefit of the invigorating sea breeze after the labours of the past few months, during which he has been the 'observed of all observers,' and the centre of attraction to the whole metropolis. There is also another reason why the animal should quit town, at least for a time, in the fact of the arrival of a-rival [sic] in the shape of the largest Tortoise in the world, who threatens to dislocate the nose of the Hippopotamus.

The third – which appeared on 15 June 1850, soon after Obaysch arrived – shows a ribbon of urgent and hurrying people winding from the top to the bottom of the page and in and out of the text until they come to a halt at a railing, behind which sits a beaming baby hippopotamus with a vaguely drawn Egyptian at his side. The fourth, also published very close to Obaysch's arrival in the Zoological Gardens, shows a smiling young hippo descending by balloon into an animated and top-hatted crowd. The balloon is

---

35 'News for the Horse Marines', *Punch*, 15 June 1850; 'Come along Fido', 6 July 1850; 'The Sea-Side Season,' 27 July, 1850; 'The Hippopotamus Arrives by Balloon', *Punch*, 12 October 1850.

decked out with the Union Jack and a naval ensign and on Obaysch's back sits a gentleman (David Mitchell?) gesturing triumphantly, with arms outspread to the appreciative multitude below. Of all the Obaysch cartoons, I find this to be the one that sums up his extraordinary presence best of all. In some ways his actual method of arrival – in the specially adapted *Ripon*, and via the crane that lifted him onto the train – was as remarkable as arrival by hot air balloon. The flags and the triumph of the human passenger demonstrate both the importance of the hippo to the Zoological Society and the specifically nationalistic inflection of the whole project: only the English could have done this. The cartoon Obaysch hangs passively and grinning inanely. He has been tamed in spite of the outrageous nature of his predicament.

The first cartoon does, of course, reinforce the notion of Obaysch as a kind of dog. The second picks up on his swimming but also, like the first, serves to socialise him within a relatively well-to-do and very English milieu as well as referring, as *Punch* had done before with regard to Obaysch's competition with baby elephants and ant-eaters, to the idea of Obaysch as a 'star' designed to pull people into the Zoological Gardens. This is not the kind of hippo who, when he gets into the water, is going to bite the other bathers in two. This image of Obaysch was carefully cultivated for the entire duration of his life in the Zoological Gardens. The third is a depiction of the huge crowds that went to see Obaysch and, I suspect, in spite of the fact that it is designed as an amusing cartoon, quite accurate.

Obaysch was undoubtedly a popular animal, and there is no doubt that his arrival transformed the fortunes of the Zoological Gardens. But can we really speak of a craze or cult? Victorians were pre-disposed, it would appear, to crazes. For example, there was the Pteridomania, which led to more or less elaborate ferneries being built on the side of every suburban villa, some projecting over the street like Puginesque Gothic air-conditioning units, to ferns in terracotta or cast iron adorning every surface capable of being decorated, and to a destruction of wild fern stocks from which some species have never recovered.[36] Or the craze for aquariums stimulated by the technological discovery of a method for maintaining aerated sea water and the aesthetic presentation of anemones and the inhabitants of rock pools in Philip Henry Gosse's *The Aquarium: An Unveiling of the Wonders of the Deep Sea* (1854) and *Tenby: A Seaside Holiday* (1856), both of which led to the notion that a tank full of weeds and fishes was an acceptable furnishing for a drawing room and to a denuding of life from English beaches that even the respite offered by the Second World War's minefields, tank traps and barbed wire entanglements was unable to restore.[37] Gosse also wrote about sea shells, so there was a related malacological craze and high prices were fetched for rarities. Many of the dealers in exotic animals – Charles Jamrach is the most prominent example – also dealt extensively in sea shells, and both scientific shell collections and objects made out of or decorated with sea shells became a material part of Victorian lay science and of Victorian popular culture.[38]

Was there really a hippo craze? The term 'hippomania' (which is more commonly, if commonly is an even vaguely appropriate word in this context, used to describe an excessive fondness for horses) was first coined to describe the cult of Obaysch by Nina Root in

---

36 On this see S. Whittingham, *Fern Fever* (London: Francis Lincoln Ltd., 2012) and D.E. Allen, *The Victorian Fern Craze* (London: Hutchinson, 1969). A. Wilkinson, in *The Passion for Pelargoniums* (Stroud: Sutton Publishing, 2007), chronicles another Victorian horticultural obsession.
37 See R. Stott, *Theatres of Glass* (London: Short Books, 2003).
38 S.P. Dance, *Shell Collecting* (London: Faber & Faber, 1966). J. Yallop, *Magpies, Squirrels and Thieves* (London: Atlantic Books, 2011) offers an interesting account of a range of other Victorian collecting fads and obsessions, as does A.C. Colley in *Wild Animal Skins in Victorian Britain* (Farnham: Ashgate, 2014).

1993.[39] I can find no evidence that it was ever used in the nineteenth century. But this doesn't mean that there wasn't a hippo craze.

I have analysed the numbers of Obaysch's visitors above. If we assume that hour-long queues and increased revenue are evidence of a craze, then we can say that there was a craze. In the first couple of years of Obaysch's life in the zoo – up to, say, 1853 – the number of articles, cartoons and poems about him in the press kept him in the public eye. This is also, arguably, evidence of a craze. There were other artifacts too. One prominent example is Louis St Mars's *Hippopotamus Polka*, which was mentioned earlier.[40] This was published in 1850 as a piece of sheet music with an attractive and amusing picture on the cover showing a demur young lady in a frothy ball gown complete with corsage being partnered by a hippopotamus in white tie and tails. The hippo is, like Oliver Hardy or Robbie Coltrane, surprisingly light on his feet. The music is undistinguished and, for a polka, rather lugubrious; the cover image is far more frequently cited than the music itself. A presumably fictitious letter (part of which was quoted in the previous chapter) describing the visit of two ladies to see Obaysch acted as a preface. What we have here is an artifact that brings together three ways of describing a hippo: a written text, a visual image (although it is a comic image, it is not a simple caricature, for the hippo's head offers a very convincing representation of the real animal), and a slightly lumbering piece of music which might well be the sort of polka that would get a hippo to take the floor. But it is really the only artifact of its kind. Everyone who writes about Obaysch mentions it in greater or lesser detail, but it is not an example of a widespread trend: it appears to have been a one-off. If we are to use the volume of secondary artifacts and images as evidence of a craze then we do not find much to support the idea of 'hippomania'.

---

39 Root, 'Victorian England's Hippomania'.
40 See Chapter 1, note 24.

There was also a play at the Haymarket, which was obviously pushed out to capitalise on the crowds pressing to see Obaysch. There is no longer any script in existence but the following contemporary review shows that it bore all the hallmarks of a rush job:

> A new extravaganza entitled The Hippopotamus has little to recommend it beyond the popularity of its title. The scene is laid in the Rosherville-gardens, where the hippopotamus is supposed to have just arrived, and the plot – a very ordinary one of conjugal jealousy – is connected with the extraordinary animal by the circumstances that the jealous husband has been irritated by his wife's frequent visits to the wonder of the day. A situation in which this husband, acted by Mr Wright, disguises himself as a vendor of fruit and ginger beer to watch the movements of his wife, and pelts the object of his jealousy (Mr Paul Bedford) with the articles of his stall, is somewhat amusing, and the hippopotamus, which is introduced towards the conclusion, is very well made up. However, the whole piece is much too loosely connected, and the combat between an uncaged lion and the more unwieldy quadruped on which the curtain falls is singularly devoid of meaning. Absurdity is of course intended, but this need not be carried out in too fanatical a spirit.[41]

However, the theatre management clearly thought that there was an audience to be found for such a flimsy effort if it was hung on the presentation of a hippo. It is a matter of great regret that no image or description survives of how the hippo was played. Was it like a gigantic pantomime horse? Or was it a huge model wheeled around

---

41 The *Northern Star and National Trades Journal*, 17 August, 1850. The Rosherville Gardens was a popular and successful zoo and amusement park in Kent which ran between 1839 and 1900 and had a very brief revival in 1936. See J. Simons, *The Tiger That Swallowed the Boy*, p. 130. The other review is to be found in *Lloyd's Illustrated Newspaper*, 18 August 1850.

2 The Several Meanings of Hippos

on casters, or a kind of gigantic puppet? Whatever was done, it is clear that the hippo was designed to look like an adult and not as a vaguely realistic representation of Obaysch himself. Another review tells us that 'The theatre was very well attended' and that at the end the hippopotamus falls dead after a fight with the lion, who then sits on his lifeless body. This is the occasion of an interesting little piece of ur-postcolonial theory:

> The moral or meaning of this we cannot pretend to fathom. It may possibly [be] intended as an allusion to our national power and pre-eminence even over the land of the hippopotamus![42]

This review also notes that:

> The hippopotamus is one of the most distinguished debutants of the season. It is, indeed, one of the 'great feats' of the year that a live hippopotamus has been brought to this country ...

In 1852 Thomas Morton and James Maddison's play *The Writing on the Wall!* (also at the Haymarket) reinforced the prominence of Obaysch in the metropolitan imagination with a weak joke in which the London Hippodrome and the London hippo are briefly conflated.

Little silver models of Obaysch are also commonly referred to in the secondary literature, as is a silver hippo-shaped pin that is supposed to have been made to the order of a young Guards' officer for his lapel. These artifacts are mentioned in almost every, if not every, description or discussion of Obaysch. The models were supposed to have been for sale as souvenirs of visits to see Obaysch. The single piece of contemporary evidence for them is, so far as I can tell, an article in *Punch*, from 31 August 1850, entitled 'The

---

42 *Lloyd's Illustrated Newspaper*, 18 August 1850. The play was also reviewed in the *Northern Star and National Trades Journal*, 17 August 1850, and *Reynold's Weekly News*, 18 August 1850.

'Hippopotamus in a New Character'. In characteristic *Punch* fashion this simultaneously promotes and debunks the idea of a hippo craze:

> We are, therefore, surprised to see him figure in a work of art, in silver, in a shop in the Strand. What his effigy is meant for – whether as an ornament for the dinner table, or a toy for the boudoir – we have no conception. We are aware he has already featured as the head of a breast-pin. A young friend (in the Guards) came to us the other day, with his coat mysteriously buttoned.
> 'What do you think I have got?' he asked, in a voice tremulous with pleasure. We avowed our ignorance. 'Look here!' he exclaimed, opening his coat, and displaying the novel bijou. 'A hippopotamus breast pin! Isn't it stunning?' And he had! The infatuated young man (who has £200 a year besides his pay, and spends £800 to our knowledge), had gone to great expense to have modeled for him a correct likeness in little of this singularly ugly animal, and was wearing it, with the pride of a discoverer, in his cravat.

I have spent many years studying, collecting, living with and, from time to time, dealing in Victorian artifacts and I have never seen one of these objects. Nor have I found any reference to one in auction catalogues or stock lists. Yet one would assume that they would have been made in reasonably large quantities and that many would have survived, especially as an antique silver hippopotamus would still be a very attractive and highly collectible item. Silver hippos – pendants, walking-stick tops and small bejeweled models – were made in Russia by several of the master craftsmen in Fabergé's St Petersburg *ateliers* later in the century, but these were not related to Obaysch in any way and were at the extreme top end of the market for luxury *objets de vertu*. They were not the relatively cheap, semi-mass-produced models of Obaysch that we might assume

2 The Several Meanings of Hippos

would have been produced for the wealthier tourist had there been a craze.

While the secondary scholarship usually speaks of models, *Punch* mentions only one item, and it may well have been an expensive centrepiece or the base for an *epergne, surtout* or *fruitier* or some such other element in the sadly departed *bric-à-brac* of the Victorian domestic table. I can believe that a young guardsman with a good credit line in Bond Street might have had a hippo stickpin made for himself, but the complete absence of little silver hippos from the early 1850s on the antique market leads me to believe that these were simply part of the comic world of *Punch* and did not actually exist except for one expensive item long since melted down or blown to pieces in the Blitz or sitting somewhere in an attic or making its periodic run through a country auction room. And in four years' time that young Guardee, if he existed, would, ironically, have boarded the *Ripon* on his way to the Crimea, taking his silver hippo with him. I wonder if he survived the Alma (very much a Guards battle) or Inkerman, or if the silver hippo was found by a Russian scavenger stripping bodies and made its way into the back-street shops of Moscow or St Petersburg, where perhaps it inspired the imagination of a Fabergé *bijoutier*.

Punch returned to the issue of hippo artifacts in September 1853.[43] At this time the Jardin des Plantes in Paris has managed to acquire a hippo of its own and *Punch* imagined Obaysch writing to him to warn him that although he might be popular now his fame would fade:

> At present you are un charmant hippopotame, the fêted curiosity of the moment; wait another twelvemonth and they will say of you, that you are nothing better than a great pig ...

---

43 'The English Hippopotamus, at the Zoological Gardens, to the French ditto, at the "Jardin des Plantes", *Punch*, 17 September 1853.

In describing the trappings of his supposedly fading celebrity Obaysch lists:

> Casts [that] were made of me in sponge cake, and adorned the pastry-cooks windows. You saw my portrait on the frontispiece of every polka. No periodical was complete without my biography, while my bulky proportions were multiplied in a thousand different shapes, either in snuff-boxes, ink-stands, salt-cellars, butter-boats, or else figured on ladies'-brooches.

As so often with *Punch* there is a grain of accuracy that causes the remainder to carry conviction. There was a polka. There was a flood of articles and images. But although I have seen nineteenth-century snuff boxes and ink stands decorated with hippos, there is nothing to connect them either stylistically or chronologically with a specific cult of Obaysch from the early 1850s.

There is, however, one statuette of Obaysch which has come down to us. This was made in 1855 by the well-known sculptor Joseph Gawen. He was a remarkable man in that he was deaf at a time when few adjustments were made for such conditions and yet he was able to carve out a successful practice with many commissions, including some for busts of royalty. The statuette of Obaysch was commissioned by Samuel Shepheard (whom we encountered in the previous chapter writing from Cairo about the capture of Obaysch) and is constructed from Nile mud. As Shepheard lived in Cairo until 1860 we must assume either that he brought this mud to Gawen on a visit to England or that it was shipped. The statuette shows a maturing Obaysch, standing foursquare on sturdy legs and with something like a smile on his face. The piece now stands in the Zoological Society of London's library and is the most tangible contemporary relic we have to testify to Obaysch's life.[44]

So, apart from all the journalism and cartoons, we have one piece of sheet music, a farce, and only very dubious evidence that

2 The Several Meanings of Hippos

there were ever more than one or two silver models. If a craze depends on a tide of consumerism, we cannot find any evidence for one in the case of Obaysch. It is true that for a while, Obaysch was the thing to see and to be seen seeing if you had the money. In that sense, we may speak of a craze. But evidence for an enduring 'hippomania' is simply not there. It is probably better to see Obaysch and hippos in general as one of the many by-waters of the bigger Victoria craze for natural history. This included not only the ferns, anemones and sea shells mentioned earlier but also fossils, dinosaurs, the acclimatisation movement (the tangential relationship of this to hippos will be dealt with in Chapter 3), exotic animals as pets, and taxidermy. This natural history craze constitutes perhaps the most comprehensive and thoroughgoing engagement with science of a broadly based public ever to have been seen in Britain, if not in any country.[45]

Although I am sceptical that the silver models of Obaysch ever existed, a model of Guy Fawkes certainly did and was mass-produced for many years. The model soldier manufacturer William Britain diversified from purely military subjects after the Great War and produced an extensive line of hollow cast-metal zoo animals including an adult hippopotamus, a young hippopotamus and a baby hippopotamus, accurately painted in tones of grey and pink. Many of these were based on real animals, as were the plastic models which replaced the metal figures in the 1950s, and the adult hippopotamus was based on Guy Fawkes. The baby hippopotamus is sitting in a position very much like that adopted by Guy Fawkes in the early *Illustrated London News* engravings of her, so I wonder if both models were actually of Guy Fawkes: the adult hippopotamus from a photo and the baby from an engraving.

---

44 This can be seen on the Zoological Society of London's website: http://bit.ly/ZSLObaysch.
45 L. Barber, *The Heyday of Natural History* (London: Jonathan Cape, 1980) offers an excellent account of this up to about 1870.

Although the hippo craze had long passed, a ghostly memory of Obaysch and his bloat thus did linger in material form in thousands of interwar British households and well into the postwar world. I kept those hippos in a box of lead zoo animals, and subsequently had the plastic ones. I played with them while Prime Minister Harold MacMillan spoke about 'the winds of change' that were blowing through Africa in 1960.

I wish I still had them.

3
A Bloat of Other European Hippos

Obaysch's story is, in part, so fascinating because of the 'first and only' status he held for three years. There was nothing else like him and so contemporary visitors to the Zoological Gardens had to find familiar metaphors to understand him. His shape could only resolve into visibility once it had been framed by comparators that established him as part of the known world and by the deployment of the various cultural and political discourses in use at the time for the purposes of organising the complex and the strange. Victorians had never seen anything like Obaysch, so to see him at all they had to find things that he was like and speak of those instead.

We are more familiar with hippos, and although we may not see live ones very often we are used to them and are surrounded by images of them and hippo-shaped artifacts. Toys, doormats, cookie jars, bookends, inflatable costumes, black crystal iPhone holders from Lalique can fill our lives if we are not careful.[1] This means

---

1 On hippos in popular culture and much else see E. Williams, *Hippopotamus* (London: Reaktion, 2017). A stimulating account of hippos in popular culture and some contemporary art may also be found in A.E. Franks, 'The Pursuit of Hippo-ness: Hippopotamus and Human', unpublished MFA thesis, Montana State University, 2014.

that it is difficult for us fully to grasp Obaysch's strangeness and understand his popularity. We construct Obaysch by means of other discourses. For us he is the victim of colonial rapacity, the representative of an endangered species that reminds us of the great extinction going on all around us, a captive in a zoo that is configured as a cruel prison, not a triumphant statement of science and education. As I have shown, all of these views were available to Obaysch's Victorian visitors and they can all be found in contemporary commentary. But they were not, as they are today, the dominant views.

In this chapter a study of other hippos, some nearly contemporary with Obaysch, others more modern, is intended to contextualise him for an audience that is far more familiar with hippos and sees them in different ways from the public of 1850. Looking at the lives of these animals and their different experiences helps us to understand Obaysch less as a singularity and more as one of a relatively small group of rare animals who were kept in European zoos (and a few farther afield), mainly but not solely in the nineteenth century. In particular the lifespan of captive hippos when compared with Obaysch's might support my view that he was always sickly. Similarly, the enormous violence of some captive hippos makes it all the more likely that such episodes were suppressed in the public record during Obaysch's lifetime.

On 20 June 1988 Knautschke the hippo died in Berlin Zoo. He was a remarkable creature who was born on 29 May 1943, which was not a good time for anyone to be in Berlin, especially a baby hippo. Like most of Berlin the zoo suffered extensive damage in the last months of the war (as did many of Germany's magnificent zoological gardens – Düsseldorf, Hamburg, Dresden, Frankfurt, Munich, Heidelberg and the Tiergarten Schönbrunn in Vienna were all more or less destroyed, animals and all) and the hippo house suffered direct damage from bombing. Unfortunately for the animals, one of the major flak-tower fortresses that were Berlin's main anti-aircraft defensive measure was located at the zoo. Knautschke's mother was killed but, amazingly, he

3 A Bloat of Other European Hippos

survived, one of only ninety-one animals out of nearly 4000 to live through the incendiary onslaught from above and the subsequent shooting on the ground as they escaped in panic through the ruined streets of a traumatised Berlin. Knautschke became a favourite character in postwar Germany and enjoyed a fame that is the most nearly like Obaysch's that I can find among the hippos of the world since 1850. He fathered many children and his death is a salutary reminder of the underlying danger of his species. He had to be put down after suffering catastrophic injuries in a fight with one of his sons, Nante. Knautschke lives on in a rather bizarre concrete statue in the derelict historic swimming pool, the Freibad Wernersee, which is currently a site of some controversy as it faces redevelopment, and in a rather more stylish life-size bronze in front of the modern hippo house at the zoo. One of his sons lives on in the taxidermy museum attached to the zoo.[2]

Perhaps there have been other hippos who, like Knautschke, have shared a fraction of the fame that was Obaysch's. One was Huberta, a South African hippo who embarked on a long trek southwards over three years and became a public figure during the late 1920s. In spite of her fame and protected status she was shot by farmers in 1931. After a trial, her killers were charged salutary fines and she was stuffed; she and can still be seen in the Amathole Museum in King William's Town.[3] Hippos were not common in South Africa – the *Bendigo Independent* reported on 6 May 1899 that the last one in Natal, who

---

2 See B. Blaszkiewitz, *Knautschke, Knut and Co.: Die Lieblingstiere der Berliner Aus Tierpark und Zoo* (Berlin: Lehmanns, 2008) [The Best-loved Animals of the Berlin Animal Park and Zoo] and D. Jarofke, *Knautschke, Knut and Co.: Das Pflusspferd Knautschke, unser Freundliche Nachbar* (Berlin: Schuling, 2012) [The Hippopotamus Knautschke, Our Friendly Neighbour]. A German friend of mine tells me that he remembers (this would have been in the late 1950s or early 1960s) his parents taking his older brother (my friend being too young) on the difficult journey from Essen to Berlin just to see Knautschke. See also G. Bruce, *Through the Lion Gate: A History of Berlin Zoo* (Oxford: Oxford University Press, 2017).

3 In fact, hippos were recognised as requiring protection in South Africa – in the Cape Colony at any rate – as early as 1886.

was estimated to be fifty years old, had been killed. The writer observed sadly but wisely that:

> All the larger and more curious creatures are disappearing at a rate so fast that this will be a monotonous world at any rate for naturalists in another century.

Another was Gustavito, a hippo who lived in the National Zoological Park in San Salvador and was an immensely popular animal. In early 2017 – while this book was being written – it was reported that the violence which mars El Salvadorean society had spilled over into the zoo and that Gustavito, who was only fifteen years old, had been beaten and stabbed to death either by drug-crazed intruders or in some obscure gang-related execution. But, following a post mortem examination, it now appears that he may have died of an illness and that zoo authorities allegedly invented the attack story to distract attention from the poor care he had received. This controversy is still rumbling on at the time of writing.[4]

The familiarity and sense of wonder that derives from longevity can help to establish a captive hippo as a favourite animal. Tanja, a hippo in Amsterdam Zoo, was put down when she was forty-nine years old, which was plenty of time for her to become established as one of the 'must sees' for visitors to that facility. In Australia, Brutus, the favourite hippo at Adelaide Zoo, is now fifty-two and still going strong but Susie, his partner, was only recently – indeed while I was writing this chapter – put down at the age of forty-nine. Brutus is the second in a line of long-lived South Australian hippos. One of his predecessors, Newsboy, lived in the zoo from 1934 to 1977 and was also much loved. But Newsboy's predecessor died young in 1929

---

4 At least forty hippos once belonging to the drug warlord Pablo Escobar now allegedly roam wild in Colombia and are beginning to pose a significant problem for the conservation authorities there. A sterilisation program is planned.

3 A Bloat of Other European Hippos

as a result of swallowing a rubber ball thrown into his pen and the very next year the hippo in Perth Zoo died of the same cause. In fact, death by intestinal strangulation caused by ingestion of a rubber ball is still remarkably common among captive hippos and I could produce a gloomy inventory of well over a dozen hippos who have died this way, some very recently. Closer to Obaysch's period, Nina, who lived in Berlin Zoo, died of a rubber ball induced blockage on 8 October 1893. According to the *Edinburgh Evening News* (9 November 1893), she died of starvation after eating nothing for four weeks. The problem for keepers is that the cause is invisible and the symptoms come on suddenly with briskly fatal results. Nevertheless, it is surprising how many hippos die young because of failures in care or security, although hippos can live to a very good age in captivity. For example, a baby hippo died at Melbourne Zoo because of the excessive heat in the summer of 1926. It is hard not to think, even though the zoo was going through a bad patch at the time and was dilapidated and under-funded, that more could have been done to provide shade and water. One must conclude that the lessons taught by Obaysch have yet to be fully learned.

The point of these little stories is that hippos have, ever since Obaysch, attracted attention and been objects of popular affection and even love. This is, I imagine, because of their rarity, because of the impact that seeing such a large animal close up has on any but the most insensitive viewer, because of the ease with which their great broad mouths can be construed as resembling a human smile, and because their natural aggression forms so little part of their public image. I also think that Obaysch himself set up a pattern of seeing hippos which has lasted to this day and that when we see, say, the dancing hippos in Walt Disney's *Fantasia* we are experiencing the resonance of the breathless excitement of the crowd that pushed and shoved its way into London Zoo to see their first ever hippo.

Although this book is specifically a biography of Obaysch, his life, as I argue above, cannot be properly understood without at least some knowledge and understanding of the other hippos that were beginning to find their way to zoos in Europe and the Americas. Bringing a hippo to Europe was bit like running a sub four-minute mile: once Roger Bannister had shown that it could be done, everyone else suddenly found that they could do it too. Once the Zoological Society had managed it with Obaysch, and then repeated the feat with Adhela almost casually, other zoos did it too. The popularity of Obaysch in England inspired a gradual accumulation of hippos in other places.

The other Everest to conquer in this regard was supervising a birth and then keeping the baby alive sufficiently long for it to pass into adolescence if not into adulthood. Much like humans at the time, once a hippo had escaped the perils of infancy then it had a reasonable chance of living to what was considered, rightly or wrongly, to be a good age, and at least into adulthood. All this had to be done without any proper knowledge of hippo behaviour, anatomy, and physiology, or of hippos' dietary and nutritional needs. We have seen how Abraham Bartlett and Frank Buckland tried hard to understand why Obaysch and Adhela behaved as they did. We have seen how Hamet's knowledge was not as well used as it might have been. We have seen how the advice of the zookeepers from Amsterdam was taken under critical scrutiny when it came to allowing Guy Fawkes to wander about Adhela's pen and swim to her heart's content. We have also seen how the death of the two baby hippos first born to Adhela enabled dissections, which did add usefully to the meagre store of knowledge that the Victorian zoologists had where hippos were concerned. As was said *à propos* of the craze for home aquaria, which ran almost concurrently with hippomania:

> At once, pet, ornament and subject for dissection, the sea anemone has a well-established popularity in the British family circle, having the advantage over the hippopotamus of being somewhat less expensive and less troublesome to keep.[5]

---

5   G.H. Lewes quoted by R. Stott, 'Darwin's Barnacles', in R. Luckhurst and J. McDonagh, *Transactions and Encounters: Science and Culture in the*

3 A Bloat of Other European Hippos

The collocation of the attributes 'pet, ornament and subject for dissection' offers an uncomfortably apt summary of the attractions of Obaysch.

Now I will examine the map of Europe and the United States between 1850 and 1900 as it if it were delineated not by towns, battlefields or ancient monuments but by hippos living in zoological gardens. This will be a highly selective rather than an exhaustive survey but I hope that it will be sufficient to give some idea of the increasing numbers of hippos that were available to the European public and the extent to which knowledge of hippos was mainly advanced through postmortems on baby hippos who had died in infancy.

In 1860 a 'tame' hippo was to be seen at the Alhambra Palace theatre. This offered a livelier spectacle than Obaysch had ever managed:

> The hippopotamus in Regent's Park is so fond of concealing every part of its person save its extremely ugly face, which usually appears on the surface of the water, like an ill-favoured bubble, that the public may fairly welcome a specimen of that rare genus forced to abide on dry land, and consequently to take that direction of its amphibious nature which it seems most anxious to avoid. A hippopotamus placed under these circumstances is now exhibited at the Alhambra Palace, where its slow walk round the arena contrasts wonderfully with the rapid feats of the Berri Brothers.[6]

Nineteenth Century (Manchester: Manchester University Press, 2002), pp. 151–181.
6   *Times*, 22 August 1860.

It appears that the hippo performed a routine that included walking round in a circle, and allowing 'a bulky gentleman of colour' to put his head into its mouth and to ride on its back while listening to a comic song about hippos. One shudders to think what was involved in its training. *The Times* professed amazement at the docility of this animal and it does seem remarkable that it could have been tamed in this way (although as we have seen, earlier hippos may have been domesticated).

The patrician tone of amused condescension adopted by the *Times* stands in marked contrast to the shriller indignation of the *Daily News*, especially when it considers the claim by the Alhambra's proprietor that the hippo is 'the only species of the pachydermous tribe that has been thoroughly domesticated':

> Perhaps allowing a very seedy and dirty Egyptian fellah to put his unpleasant-looking head in your mouth, is to be 'thoroughly domesticated;' perhaps to walk very slowly and surprisedly round in a circle is an 'interesting performance,' and not to revenge yourself on the keeper who bestrides your back and whips you heartily is to 'exhibit great docility.' Certainly this is all that is done by the new-comer, a young and diminutive specimen of the hippopotamus, who has much promised for him, but who does not perform at all. Those wishing to see a fine hippopotamus will see two at the Zoological Gardens. Those wishing to see a performing hippopotamus will not see one at the Alhambra.[7]

This article also tells us that the proprietor gave £3000 for this hippo, which was sold to him by John Petherick, the consul for the Sudan. This is not accurate. Petherick did eventually sell the hippo, but not to the Alhambra.

This is a very interesting snippet, as Petherick did import a hippo into England in 1860 (this is dealt with above) and the nature

---

7   *Daily News*, 21 August 1860.

3 A Bloat of Other European Hippos

of its capture and transportation was described at length by Frank Buckland. This hippo briefly lived in Regent's Park and then, under the name of Bucheet, spent time at Hull Zoo before being exported to Canada and then to the USA, where he ended his life as one of Barnum's circus animals. The evidence surrounding Bucheet makes him seem a Protean beast and it would be easy to conclude that there were at least two hippos or maybe even three circulating in England and subsequently across the Atlantic in 1860. But when all the strands are drawn together I am sure that there was only one, and that the somewhat wretched animal exhibited at the Alhambra Palace was actually Bucheet. I believe his co-performer was his Arab guardian Salama, who subsequently turned up in Barnum's Greatest Show on Earth. My guess is that having been frustrated in his attempt to sell Bucheet to the Zoological Society, and perhaps still smarting from the loss of the three other hippos he had captured at the same time, Petherick leased the animal to the Alhambra while he negotiated a sale with the American showmen. I will return to Bucheet later in this chapter.[8]

In 1878 a hippo was to be seen at the aquarium in Edinburgh. It had come to Scotland via an easier route than that available to Obaysch and Adhela, as it travelled up the Suez Canal. It had been captured on the White Nile near Kassala and was a maturing male.[9] Another arrived in Liverpool on the steamer *Calabar* in 1897. This was a small animal of undetermined gender and about three months

---

8   The issue of Bucheet, the performing hippo and the various hippos that pop up in the USA and Canada at much the same time was a vexatious problem in the researching of this book. Sometimes it appeared that there were two or three different animals but there was, in fact, only one. R. Reynolds III's paper 'America's First Hippo' (*Association of Zoos and Aquariums Regional Conference Proceedings*, 1996, pp. 346–351) offers a full account of this animal. Hull City Council has recently commissioned a statue of Bucheet which stands where the old zoo used to be; although he had a very unhappy life, he is at least now memorialised.

9   'Hippopotamus at the Edinburgh Aquarium', *Fishing Gazette*, 20 December 1878.

old. It was from West Africa but the narrative of its capture reads not unlike that of Obaysch:

> The capture of the hippopotamus was brought about in an ingenious manner. It seems that it was born in the upper reaches of the Gambia River, West Africa. It is customary for the male to eat the young [this is not true!]; and to prevent this being done, the mother usually secretes her offspring. The present infant was placed by the mother in a hole dug in the banks of the river, being afterwards covered by weeds and grass. This operation was watched by the natives in the village adjacent, and when the mother went to the opposite side of the river, they pounced on and secured their youthful prize. This was done by covering the baby with a net and securing it with ropes. They brought their capture with all possible speed to the village and sold it to a trader who was a passenger on the Calabar.[10]

The interesting thing about this account is that it suggests that the natives kidnapped the baby hippo as a speculative venture, anticipating that they would be able to sell it to a European trader. Whether this was done to order or was pure speculation is an open question, but their entrepreneurial impulse paid off and they made the sale.[11] My guess is that they were used to selling animals in this way and, certainly, the trader who bought it must have had some preparations in hand to transport his new creature up the river and then ship it all the way to Liverpool. Unfortunately I cannot trace what happened to it on arrival but my guess is that the trader had it in mind to sell it to one of the big Liverpool animal dealers – probably William Cross, who had sufficient facilities to offer a rhinoceros for sale in 1904 so could easily have handled

---

10 'A Baby Hippo', *Young England*, n.d. [1897?].
11 In fact at Entebbe British officials would even accept wild animals in lieu of tax, with 300 rupees the going rate for a hippo. See H. Ritvo, *The Animal Estate* (London: Penguin, 1990), p. 247.

3 A Bloat of Other European Hippos

a baby hippo. The Liverpool dealers had achieved pre-eminence over their London counterparts since the opening of the Suez Canal had shifted the balance of the exotic animal trade in favour of continental (especially German and Belgian) merchants, and left Liverpool as the only English port where ships carrying exotic animals – especially from South America – were regularly docking in the way they had in London up until the 1860s.[12]

This hippo was lucky. Only a year later the *Calabar* foundered off the coast of Liberia. so it may have joined the ranks of the many animals who died in transit. If the ship's captain tolerated animals on board then, given the money at stake, it seems unlikely that there wasn't at least one animal penned up somewhere on the ship whenever it made its regular voyage between West Africa and Liverpool. By 1907 the intrepid American journalist Richard Harding Davis was hunting hippos in West Africa and described his unsuccessful attempts to shoot one.[13] His account is interesting as it shows how relatively quickly this region had developed from a place of chaos defined largely by the slave trade to a point where an American gentleman could go on a hippo hunting party. Of course, this development held terrible consequences for the long-term survival of hippos. Given the size of the *Calabar* hippo and the location of its capture there is also an outside chance that it was actually a pygmy hippo, possibly the almost mythical *Choreopsis heslopi*. I think it unlikely that any European involved in the capture and transportation of this animal would have been able to tell the difference between a hippo and pygmy hippo, especially when the creature concerned was so young.

---

12 See J. Simons, *The Tiger That Swallowed the Boy* and 'The Scramble for Elephants'. An examination of Cross's advertisements and stocklists for the period shows a predominance of birds and animals from South America. This is a symptom of the decline in the wild animal trade in England more generally as more and more African and Asian animals made their first landfall on the continent and were snapped up by the dealers and zoos there.
13 See R.H. Davis, *The Congo and Coasts of Africa* (New York: Charles Scribner's Sons, 1907).

The most unfortunate hippo in nineteenth-century England, however, was the one who perished in the massive fire that destroyed a large part of the Crystal Palace in 1867. The remainder of the palace went up in smoke in 1936 (as indeed did two replica Crystal Palaces: the New York Crystal Palace burned down in 1858 and the Munich *Glaspalast* in 1931. It is surprising that glass is so flammable). It is often forgotten that the building where the hippo was living was the remnant of the much larger original and also that when that original was rebuilt at Sydenham it was put together in a different configuration from what was originally laid out in Hyde Park. By 1867 the building was being used as what we now called an exhibition centre and this included a small menagerie. This, alas, was placed in the area where the fire (caused by a gas leak) took hold and the baby hippo who lived there was burned to death. Some animals were saved, including a parrot who was carried to safety by no less a saviour than the Duke of Sutherland (who just happened to be passing, I suppose) and bit his hand to show his gratitude, according to the *Spectator* (5 January 1867). The hippo had been born in Amsterdam and was bought for £800 with a view to selling it on to the USA. It was given a post-mortem by the zoologist Edward Crisp. He was the first scientist to conduct systematic and thorough research into the relationship between captivity and early mortality so is significant in zoo history. Crisp's operation yielded a fine description of the animal's gut anatomy. Crisp couldn't resist popping a morsel of the dead hippo into his mouth; he pronounced it delicious and 'whiter than any veal I have ever seen.'[14] Perhaps the most important thing to note about this hippo is that it had been

---

14   Quoted in S.K. Etringham, *The Hippos* (London: Bloomsbury, 1999), p. 23. In the same year Gratiolet dissected two hippos at the Jardin des Plantes. Fourteen years later, H. Chapman offered another important contribution to the understanding of hippopotamus anatomy in 'Observations from the Hippopotamus', *Proceedings of the Academy of Natural Sciences of Philadelphia* 3 (1881), pp. 126–148.

3 A Bloat of Other European Hippos

born and survived in captivity. I believe that it was the first hippo to have been successfully raised in this way in Europe.

The foundation of the Zoological Gardens in Regent's Park was motivated not only by a desire to establish a first-class scientific facility in London but also, more nationalistically, to catch up with the French. When it came to zoos France had got there first and Le Ménagerie du Jardin des Plantes opened in Paris in 1794 as a showcase for revolutionary arts and sciences and as a home for the few remaining animals from the royal collection at Versailles. These were subsequently augmented by animals stolen to order by French armies as they roamed over Europe and by collections brought back by French voyages of exploration, especially into the Pacific and Australasia. To build the collection – beyond the few animals left after the destruction of the royal menagerie at Versailles and other aristocratic menageries such as that at Chantilly – the government also abolished all of the street menageries and animal performances that were common in Paris at the time and rounded up the animals.

Although the Jardin may be characterised as the oldest scientific zoo, the Vienna Zoo, which was founded in 1752 as the Tiergarten Schönbrunn, an imperial menagerie, is the oldest continuously operating public zoo of a recognisably modern kind. But the Jardin was undoubtedly the first public zoo of a scientific character in Europe.[15] Although its construction had been approved in 1792, it was not until two years later that an established facility was set up. Before that there was a display of animals but not in a permanent and purpose-built facility. Its foundation created

---

15 See R.W. Burkhardt Jr, 'Constructing the Zoo: Science, Society and Animal Nature at the Paris Menagerie 1794–1838' in Henniger-Voss, *Animals in Human History*, pp. 231–257. Given the importance of the Jardin it is surprising that a really good history is yet to be written. The build-up to the foundation has, however, been well covered by E.C. Spary in *Utopia's Garden* (Chicago: University of Chicago Press, 2000) and more recently in the dazzling set of studies presented in P. Sahlins, *1668: The Year of the Animal in France* (New York: Zone Books, 2017).

a spirit of competition, sometimes friendly, sometimes not so friendly. In the late 1820s there was a spate of giraffe rivalry, with both the London and Paris zoos (and the Vienna Tiergarten) sparking 'giraffomania' as they vied for the most spectacular success with their giraffe display. The French won – their giraffe lasted eighteen years while the London animal survived for only two before it went to John Gould to be stuffed.[16] Given this history, it is not surprising that the Paris zoo is where we should look for the first hippo, after Obaysch, to come to Europe.

The animal was called Coco and arrived in the Jardin des Plantes in 1853. His age was estimated at thirteen months. As mentioned above, *Punch* (17 September 1853) published a letter purporting to be from Obaysch, signing off as 'your old camarade de Nil', advising the new arrival not to depend on the longevity of his current fame. Coco had a very similar early history to Obaysch and was looked after for eight months by the French consul, who kept him in his country house until suitable shipping was arranged and until the hippo had gathered strength for the journey. Like Obaysch he had an Egyptian keeper. An image published in *Chatterbox* (27 July 1870, but showing what is clearly a very young hippo, so presumably recycled from a much earlier image) shows Coco standing in a pool with a man in Oriental dress looking down on him from behind some railings. Early accounts of his presence in the Jardin des Plantes are often unflattering, with this from *La Nouvelliste* (1 November 1853) a typical example:

---

16 Allin, *Zarafa*. In fact a comparison of the records of the lifespans of animal in French and British collections suggests that the French were much better zookeepers than the British at this time. For example, while wombats in early nineteenth-century England were lucky to struggle on for more than a couple of years, those that arrived in the Jardin des Plantes at much the same time lived into their early teens. See J. Simons, *Rossetti's Wombat*, pp. 19–25.

3 A Bloat of Other European Hippos

Coco – il s'appelle Coco – n'a rien de gracieux, et son physique est des plus désobligeants, il a l'oeil stupide, le galbe ignoble, la taille ramassé ...

[Coco – his name is Coco – has nothing gracious about him and his physique is most offensive, he has a stupid eye, a horrible form, a squat height.]

La Presse (10 April 1854) referred to Coco as the first hippo to have been seen in Europe since Roman times. This is an interesting deployment of the concept of Europe given that everyone knew that there were hippos in England. I suppose anyone wanting to understand the history of recent upheavals in the European Union and the British withdrawal need only contemplate this casual assumption. The English newspaper the *Leader* somewhat huffily pointed out that:

> 'Coco' as the hippopotamus is called, has been for weeks past an occasional aliment to the Charivari. He is regarded with the same affectionate interest that attended his cousin in Regent's Park, and the tip of his nose is anxiously looked for by crowds of morning visitors. The Parisians are taught to believe that their hippopotamus is the first specimen of the race ever brought to Europe, totally ignoring our earlier acquisition.[17]

A second hippo, a female who was called Bichette, arrived in France in 1855, completing the diplomatic circle and putting France back on a par with England when it came both to hippo importation and relationships with Egypt. The next year's issue of *La Guide des Étrangers dans la Museum d'Histoire Naturelle* proudly stressed its possession of:

---

17   *Leader*, 3 September 1853.

Deux jeunes hippopotames du Nil-Blanc, mâle et femelle, arrivés successivement de l'Egypte, et donnés à l'Empereur des Français par le vice-roi d'Egypte et par son frère.

[Two young hippos from the White Nile, a male and female, have arrived successively from Egypt, and have been given to the Emperor of the French by the viceroy of Egypt and by his brother.]

Notice that Coco and Bichette are seen as gifts directly to Napoleon III and not, as in the case of Obaysch and Adhela, to a scientific establishment or learned society.

There were several babies born from this mating pair. However, the keepers of the Jardin des Plantes did not initially have great success in keeping the young hippos alive and between 1858 and 1867 it appears that as many as nine were born, eight of whom died. If nothing else their bodies provided useful specimens for the continued examination of hippo anatomy.

On 17 May 1858 Frank Buckland reported in a letter to the *Times* (published 18 May) that a baby born on 10 May of that year had died. It was an interesting birth as it happened underwater and, if Buckland is to be believed, the baby simply bobbed up at six in the morning on the day of its birth. It swam about for while and then tried to get out of the water but because of the design of the pool (no slope) it could not. Its mother attempted to help it but in the course of assisting it up the step she managed to injure it so badly that it died later that day. This was, Buckland thought (rightly I believe) the first live birth of a hippo to have taken place in Europe at least in modern times. He sadly opined that 'we cannot but regret the untimely end of the little hippopotamus.'

In 1859 a similar event happened and, again, a baby was born under water. This little one survived for a few days but was then killed by its mother. The interesting thing about this event is that reports of it suggest that the death of the baby hippo reported by Buckland in 1858 was not quite as innocent as he thought:

On the former occasion the female hippopotamus repulsed her
offspring, would never let it suck or come near her, and in pushing
it away violently inflicted a wound which caused its death.[18]

Instead of trying to help her baby from the water it appears that the Parisian hippo was actually viciously rejecting it. It is noteworthy that although the circumstances of Adhela's early disastrous attempts at motherhood appear very similar to Bichette's and had the same sad ending, her aggression was not seen, in public at least, as the reason for the deaths of her babies. This general tendency among the London zoologists to conceal or diminish a hippo's propensity to violence is not found when they speak of French hippos. And, indeed, the French zoologists are much more frank about the realities of keeping aggressive animals like hippos and the destruction they can wreak. While the British zoologists were playing down the violence of their charges the French zoologists were painting a slightly different picture based on the realities of the situation (especially given the terrible record of dead hippo babies in the Jardin des Plantes):

> À Londres, le naturel violence du male a toujours empeché de le réunir à la femelle. À Paris, les deux individus ont dû, de meme, rester quelques temps séparés.

> [In London, the natural violence of the male has always prevented his being brought together with the female. In Paris the two individuals must likewise spend some time separated.]

The comment about the violence of the male (Obaysch) is interesting. Did the French zookeepers, who would have had regular contact with Bartlett at this time, have access to a different narrative than that given to the British public?

---

18   Buckland, letter to the *Times*.

This time, however, the maternal bond appeared to be forming and the baby was allowed to suckle and to ride on its mother's back. It stayed mainly in the water but also got out and walked around the pen (either it could manage the step or they had put a ramp into the pond). It was visibly putting on weight. But then things took a turn for the worse:

> During the night the mother was seized with a fit of rage and suddenly attacked it. 'It is an extraordinary fact,' says M.I.G. St Hilaire, 'that the females of these mammiferous animals abandon their young, ill treat, and even devour them. But it is almost without example that when the mother has adopted the young one and given it suck, it should do so. It is true however that there is no animal more irascible and brutal than the hippopotamus.' The event having occurred under water and in the night, the keeper was not able to give a full account of what took place, but the results are but too clear. The mother must have seized the young one by the stomach in her formidable jaws, as five deep marks of her teeth are visible, and she must have attacked it with her tusk, which pierced the left breast into the lungs.[19]

The newspaper *L'Echo du Parlement* reported on this tragic event and put an interesting political and darkly anthropomorphic spin on the event:

> *Peut-être cette mère hippopotame a t-elle cédé à un mouvement patriotique, et a t-elle donné la mort à son enfant plutôt que de l'élever pour la captivité ... l'hippopotame, qui tue ses enfants pour les delivrer.*
>
> [Perhaps this mother hippopotamus has given in to a patriotic urge and has given death to her child rather than raise it for

---

19 Buckland, letter to the *Times*.

captivity ... the hippopotamus which kills its children in order to free them.][20]

The newspaper speculates on which is the more savage: the infanticidal hippo or the scientists who keep her in slave-like captivity. This might seem a particularly French line of speculation but, in fact, at least some reactions to the Zoological Gardens contemporary with Obaysch made a similar point like this from the *Quarterly Review* in 1855:

> Why do we coop these noble animals in such nutshells of cages? What a miserable sight – to see them pace backwards and forwards in their box-like dens.[21]

After the infant's death in 1859 it was decided that when the hippos bred again the newborn baby would be taken from the mother immediately. This plan cannot have worked given that Bichette's subsequent offspring also seem to have met their deaths as a result of her attacks on them. It is interesting to note how little use appears to have been made of the experiences in the Jardin des Plantes when it came to understanding Adhela's behaviour with her first two babies. This may have reflected the tendency of the Zoological Society to minimise and even misrepresent the aggressive nature of its hippos, but the scientists would not have been ignorant of what had happened in Paris, and this may explain the anxiety Bartlett felt when little Guy Fawkes disappeared under water for so long or had trouble getting out the pool. But when Adhela seemed unable to feed

---

20 *L'Echo du Parlement*, 25 July 1859.
21 Wynter, *Curiosities of Civilisation*, and also quoted by A.N. Wilson, *The Victorians* (London: Hutchinson, 2007), p. 79. We make a mistake to assume that the Victorian attitude to progress and empire was a single shout of triumph. Kipling's 'Recessional' ought to remind us of this but we forget it all the time.

her first baby no reference is made to the risk of an attack using the accounts from Paris.

A third baby died in 1860 – notice that as with Adhela, Bichette's breeding frequency at this point appears to have been an accelerated version of the rhythm found in the wild. Of this unfortunate animal *Le Journal Amusant* opined, perhaps with a satirical nod to the efforts of the zoo keepers and their explanations of Bichette's behaviour, that:

> *L'hippopotame vient de manger son fils dans un accès de vive tendresse*
>
> [The hippopotamus comes to eat its son in a burst of deep tenderness.][22]

A fourth baby died in 1862, a fifth in 1866 (noted in *Le Petit Journal* as the eighth to have been killed by Bichette, which, if true, is remarkable) and a sixth in 1867, all apparently killed by Bichette. A baby who was born in 1865 may have survived. Buckland records this and 'believed' that it had survived because it was 'taken from the mother immediately at birth'.[23] I believe however that this baby also died at some point in its infancy, as in recording the birth of Guy Fawkes to Adhela in 1872 *Le Journal Officiel de la République Française* enumerated the hippos who had been born in Europe to that date (six in Amsterdam, two in Paris and three, including Guy Fawkes, in London) and commented that, so far, every one had died in infancy (although one of these was the hippo killed in the Crystal Palace fire). This account, which is surely trustworthy given the status of the journal and corresponds with figures given by Frank Buckland in the *Morning Post* (6 November 1872), markedly

---

22 *Le Journal Amusant*, 19 May 1860.
23 F. Buckland, *Curiosities of Natural History*, 3rd series (London: R. Bentley, 1868), p. 114.

3 A Bloat of Other European Hippos

conflicts with other newspaper reports on the births of babies to Bichette.

Two hippos were recorded as living in the Jardin des Plantes in 1887: one was Bichette, and I have yet to determine the other's name. We learn of it solely because on 5 January it killed one of its keepers. He was a man called Braner and was cleaning out the hippo enclosure when the hippo caught him in its mouth and severed his carotid artery.[24]

In 1894 the English journalist Henry Sutherland Edwards visited the old Bichette. She was still alive and so was her reputation as a killer. Sutherland noted:

> A striking peculiarity of the female hippopotamus in the Jardin des Plantes is that she has given birth several times to a tough-skinned baby and always, or nearly always, killed it immediately with her terrible teeth.[25]

We have seen that none of the babies survived although at least one enjoyed a few days' maternal love before the rage set in again. It is surely noteworthy that after nearly forty years this vicious creature was still living in the Jardin des Plantes, a specimen as much of the inherent violence and aggression of her species as of anything else. In 1890 *Le Petit Parisien* ran a piece on the oldest animal in the zoo, which was, of course, Bichette:

> *Jusqu'à l'hiver dernier, ce 'doyen' avait joui d'une florissante santé; aujourdhui il commence à connaître les infirmités de l'âge et il est vraiment hideux.*

---

24 'Killed by a Hippopotamus', *Daily Gazette for Middlesborough*, 7 January 1887.
25 H. Sutherland Edwards, *Old and New Paris* (London: Cassell, 1892–1894), p. 152.

> [Until last winter this doyen had enjoyed a flourishing state of health; today it begins to know the infirmities of age and is truly hideous.]²⁶

The 1890 issue of *Causeries Scientifiques* (published in 1891) confirmed that Bichette had, until the onset of the cold, been in robust health:

> *L'hippopotame qui depuis 36 ans était en excellente santé a eu la peau entamée par des fissures profondes.*
>
> [The hippopotamus which has been in excellent health for thirty-six years has had its skin opened by deep cracks.]

If 'Strange Patients', a report on veterinary medicine at the Jardin des Plantes in the *Friendly Companion and Illustrated Instructor* (1 May 1892), is to be believed Bichette's infirmities appear to have been chilblains:

> There is also a hippopotamus which suffers from chilblain, left by last winter's frost. The wounded places are rubbed with Vaseline and the enormous creature shows its gratitude by uttering comfortable grunts.

In fact, the winter of 1891 was so cold that some of the Jardin's animals froze to death and Bichette herself had to be helped by the provision of a hot water bath. The baby Baptiste is also reported as suffering from chilblains at this time. This cracking and the ulceration caused by chilblains are precisely the same ailments from which Obaysch suffered. The French zookeepers would have had access to Garrod's post-mortem report and this may have stimulated them to give Bichette some hotter water.

---

26   *Le Petit Parisien*, 24 October 1890.

3 A Bloat of Other European Hippos

Bichette died on 4 February 1897, aged at least forty-three – she had lived to the good age. In fact, she died in the very bath that may have saved her life during the hard winter of 1891. (*Le Figaro* noted on 5 February,1897 that '*L'hippopotame femelle de Jardin des Plantes est mort hier soir dans son bain*' – 'the female hippopotamus in the Jardin des Plantes died in its bath yesterday evening'). She was a subject of national mourning:

> Unusual excitement prevailed today in the Botanical Gardens, whither people flocked in the dry mild weather to attend what may be called the funeral of a hippopotamus. The deceased river horse has a history and is not to be buried or cremated, but was removed this morning to the animals' Morgue in the Museum of Natural History. There a post-mortem examination is to take place, and the huge beast will subsequently be preserved in stuffed, skeletonic form, like the big whales in the Cour de la Baleine. The hippopotamus was captured young, in the Upper Nile, so far back as 1855, and was sent to France in charge of a Nubian expert. The animal, which was forty-eight years old, was found dead in its big bath yesterday, and was lifted out with great difficulty by the attendants ... The loss of the old hippopotamus is deeply deplored by all visitors to the Jardin des Plantes as well as by the museum officials, and it is no wonder that the animal has received long obituary notices in the newspapers.[27]

It may be noted that the history of this animal is not unlike that of Obaysch, Nubian expert and all, and it is clear that the French zoologists had learned from the Zoological Society's method and experience when it came to the task of transporting live hippos from Africa. France did not really have a national press on the English model and its newspapers, at least the Parisian ones, were frequently associated with political parties, movements or tendencies. This

---

27 'Paris Day by Day', *Daily Telegraph*, 5 February 1897.

meant that while hippo matters were reported, the French press never whipped up anything like the cult of hippomania (real or imagined) that swirled around Obaysch and Adhela. Although a comic piece in *Le Figaro* on 9 September 1858 has a reference to 'la polka de l'hippopotame', which is presumably the same piece that had been popular in England in 1850. French hippos are rarely named in such newspaper articles, so it is occasionally difficult to work out which hippo is being referred to. Indeed one of the fullest early accounts of the Paris hippos is to be found on 14 May 1858 in the Belgian newspaper *L'Independance Belge*.

Perhaps the fact that Coco showed the challenges of keeping hippos very early in his captivity made people less fond of him than they were of Obaysch. The *Sydney Morning Herald* on 10 March 1854 reported an unfortunate incident:

> A strange scene took place the other day in the Jardin des Plantes. Amongst the persons collected round the enclosure of the hippopotamus was an elegantly dressed lady, accompanied by a little King Charles' dog. The little animal having gone inside the rails of the paddock, was at once seized by the hippopotamus and swallowed in an instant.

It will be remembered that when Obaysch did something similar the incident was suppressed. An article published in the New Orleans paper the *Thibodaux Sentinel* on 2 January 1897 claimed that Coco shared an enclosure with an elephant who pulled his ears with his trunk and stood over Coco's food so that he couldn't get at it. I doubt if this is true, but, if it is, it might help to account for Coco's irascibility.

Coco and Bichette lived precariously through the siege of Paris (although some sources say there were three hippos in the Jardin des Plantes at that time, the overwhelming probability is that there were just the two – Coco and Bichette). As the siege progressed and the situation in the city grew more desperate most of the animals were auctioned for meat. But the hippos were both passed in. This was partly because the opening price of 80,000 francs was so steep and partly because, as *Le Figaro* put it, with a magnificent disregard for the tragic circumstances unfolding in the French capital at the time, '*la viande doit être mediocre*' (the meat must be mediocre).[28]

Coco died on 12 August 1873; a note mentioning the death of a hippo appeared in the *Le Petit Journal* on the thirteenth. This was after an illness mentioned in *La Presse* on 9 August 1873. The illness is noted as nostalgia for the Ganges and on 13 August the same paper reported that '*l'enorme pachyderme a rendu son âme à Brahma*' ('the enormous pachyderm has given up its soul to Brahma') – *La Presse* seems to want to believe that hippos came from India. Coco isn't named at any point, which shows the difference in hippomania between France and England. He was about twenty-two when he died. As we have seen, this is no great age for a hippo. Loisel says he lived until twenty-seven but this is surely wrong.[29]

Coco did have one marvellous adventure: he escaped in July 1870 and swam down the Seine, frightening the passengers on a steam launch, shaking a washing barge, much to the dismay of the washerwomen on board, and clearing the baths beside the Pont d'Austerlitz. He had been taken to the river in a cart every day as the hot summer and consequent drought had dried up the pools in the Jardin des Plantes. This day Coco broke his chain and was only recaptured after a thrilling chase involving a flotilla of fifty small boats as well as a veritable gala of swimming zoo keepers, one of whom bravely, if foolishly, actually managed to clamber on top of him. The English journal *Illustrated Police News* (23 July 1870) published a wonderful engraving showing him joyfully upsetting a boat and being watched anxiously by a crowd on the bank. It is

---

28 *Le Figaro*, 2 January 1871.
29 It grieves me to point this out as I think Loisel is the finest of all zoo historians. All subsequent workers in the field should be in his debt. It is a pity that his magisterial work has never been translated into English.

noteworthy that only a male hippo was taken to the river. Bichette was simply too unpredictable to be handled in this way.[30]

After Coco's death the Jardin acquired other hippos, and one was also obtained in July 1896 by Paris's other zoological establishment, the Jardin d'Acclimatation. The price of hippos had decreased in the last quarter of the nineteenth century, showing perhaps that ease of capture and transportation had compensated for the growing rarity of the animals, and that zoos were becoming well-stocked with their own hippos. In 1877 the biggest clearing-house for exotic animals was at Antwerp and 50,000 francs was the going rate for a hippo. By 1893 the Antwerp dealers were offering a six-month-old hippo for 12,000 francs but the best offer they could get was 5000.[31] When Bichette died the *Daily Telegraph* reported her value at £1200 pounds and noted that Alphonse Milne-Edwards (the current director of the Jardin) had recently bought another hippo (probably the one known as Kako) for only £400. In 1896 a new hippo, called Baptiste, was born; the *Sydney Evening News* (13 February 1897) reported that his mother had died in childbirth – a reversal of the usual pattern – but this shows that there was, at some point, at least one other mature female keeping Bichette company. Perhaps this was the hippo who killed the unfortunate Monsieur Braner in 1887.

In 1897 there were four hippos remaining in residence at the Jardin des Plantes following the death of Bichette. These were Antoine (who was three years old and was apparently sickly: on 12 August 1898 *Le Petit Parisien* reported him as ill), Louisette (also known as Lisa), who was just one year old (she was purchased from Antwerp and there seems to have some intention to mate her with the animal in the Jardin d'Acclimatation), Baptiste (his name is recorded by the Belgian newspaper *Le Vingtième Siècle* and the *Sydney Evening News* but is not found in any French source

---

30 See also 'A Hippopotamus in the Seine', *Tamworth Herald*, 16 July 1870.
31 *Le Figaro*, 23 November 1877 and 15 January 1893.

that I could discover) and Kako, who was probably acquired at the age of just fourteen months in August 1896 and was subsequently mated with Louisette.[32] Kako was one of a new wave of hippos and reflected the changes in the imperial world map. He came directly from the French colony of Senegal rather than from Egypt.

In July 1898 there is a report that:

> the venerable hippopotamus at the Jardin des Plantes is very seriously ill ... The dear beast (la chère bête) passed a very restless night, but is slightly better this morning. Its grunts and groans disturbing its neighbours' rest, it has been deemed advisable to remove it to a more isolated cage. This enormous animal, the largest in Europe, has been now nearly twenty years an inhabitant of the French capital, and its impending demise is evidently looked on as a national calamity.

This notice, entitled 'A Parisian Favourite Dying,' comes from the *English Sunderland Daily Echo and Shipping Gazette* (29 July 1898) and I have not been able to corroborate it with any French newspaper. Nor have I found evidence of the acquisition of a hippo in or around 1878. It is another part of the puzzle of the French hippos. My conclusion is that it can only be a very belated report on the death of Bichette.

Kako was known as '*le terrible*' and '*le misanthrope*' for good reason. In 1901 he killed his keeper Monsieur Landy and in 1903 he killed Landy's successor, the experienced zookeeper Jean-Baptiste Lancelle, who had worked in the Jardin for thirty-one years and at the time of his death was its longest serving employee. A soldier of the colonial infantry was passing and attempted to rescue Lancelle by stabbing at Kako with his sword-bayonet but the hippo would not release Lancelle, who was dead by the time rescuers arrived.

---

32 The suggestion that Louisette was to be mated with the hippo in Le Jardin d'Acclimatation is found in the *Bulletin du Muséum d'Histoire Naturelle*, 25 May 1897.

Le Petit Journal noted that Lancelle's effects were removed from the enclosure '*sous l'oeil impassible de Kako, qui ne semble montrer aucun remord pour son crime*' ('under the impassive gaze of Kako, who seemed not to show the slightest remorse for his crime'). At the time, you could buy a postcard of Kako with the caption: '*Kako, dit le Terrible (victimes Landy 16/9/01; Lancelle 1/7/03) et Lisa, sa compagne.*' A children's book about Kako puts the day of the attack on Lancelle as 15 July and suggests that Kako was in a fragile mood because he had been kept awake all night and frightened by the many fireworks set off for Bastille Day.[33] This is a nice idea but poor Lancelle actually met his doom on the first of July, so Kako has no mitigation beyond the 'crime passionel' consequent on his natural propensity to violence.

One episode which would have disturbed the life of Obaysch in a similar way was the explosion of a barge that was carrying gunpowder down the canal which runs past Regent's Park on 2 October 1874. This caused massive damage (it was most probably the biggest explosion to have happened in London until the Blitz) and damaged enclosures in the zoo. Some reports say that animals were killed but there was certainly great anxiety that animals would escape and a troop of the Horse Guards were brought into the zoo just in case. Accounts do agree that the animals set up a huge commotion and one can imagine that Obaysch and Adhela played their part in that. One notable casualty of this explosion was the ceramic and glass collection of the artist Lawrence (later Sir Lawrence) Alma Tadema. His beautiful aesthetic mansion, Townsend House, stood adjacent to the canal. The barge was carrying hazelnuts as well as gunpowder and Alma Tadema got a broadside.

---

33   E. Pollack, *Kako Le Terrible* (Paris: La Joie de Lire, 2013). In November 1999, the director of the zoo at Pessac was killed by a hippo who was apparently thrown into a jealous rage when he saw the director riding a small tractor (*La Dépeche*, 2 November 1999).

3 A Bloat of Other European Hippos

This completes the survey of French hippos. Other hippos were kept in Berlin, where the first hippo house was built in 1885; Hamburg; Antwerp, and Budapest from the 1870s.[34] The first two arrived in Amsterdam in 1860. This was a pair called Hermann and Dorothea – people had a more literary turn of mind in those days.[35] A baby was born at Amsterdam in August 1862 and was doing well, being looked after by its mother without any problems. However, the father, who was in the next-door enclosure, became increasingly agitated and eventually tried to break down the partition between himself and the other two animals. The more violent he became the more dangerous to the baby were the mother's attempts to keep it away from the dividing wall, to the point where the keepers feared for its life and removed it. Unfortunately it died the next day, either through accidental injuries it had received from its mother or through the sudden change of diet necessitated by its removal from her.[36] In 1870 Amsterdam had the rare event of hippo twins. On 4 March 1896 the Queensland newspaper the *Warwick Examiner and Times* reported that a hippo living in Hagenbeck's *Tiergarten* in Hamburg had a rude awakening one night when a kangaroo climbed over the fences separating its enclosure from the hippo and set about him, 'playing a tattoo on the most vulnerable portion of his adversary's huge carcase' and nimbly avoiding his snapping jaws. A keeper heard the commotion and, lassooing the kangaroo, dragged him 'ignominiously' away.

This is a survey mainly of European hippos, but there is still a circle to complete regarding Bucheet. After his brief stay in Hull

---

34 On Antwerp hippos see *Le Journal de Bruxelles* (5 March 1859), *Le Peuple* (27 May 1886) and *L'Indépendance Belge* (17 October 1887). By 1907 Amsterdam and Antwerp were seen as the places where the most expertise and best success in hippo breeding were to be found. See E. Velvin, *Wild Animal Celebrities* (New York: Moffatt, Yard & Co., 1907), p. 90.
35 The best account of these hippos I have found is in the German periodical *Westermanns Jahrbuch und Monatshefte der Illustrierte Deutsche* (Braunschweig: G. Westermann, 1872), p. 284
36 *Western Flying Post; or, Sherborne and Yeovil Mercury*, 5 August 1862.

he was sold to the American showman G.C. Quick for a reported $40,000 (although the American circus owners invariably inflated the prices that they claimed to have paid for things) and left Liverpool on the *City of Manchester* on 3 October 1860. He landed in New York on 19 October and was then exhibited at Spalding and Rogers's Museum in New Orleans. The next year he was on show in Havana as 'Quick's Colossal Hippopotamus' and apparently earning $3000 per week. Later in 1861 he was purchased by Barnum, who claimed him as the first hippopotamus to have been brought to the USA and the first exhibited there. This was true, but it had not been done by Barnum, as he implied. Although the rival showman and menagerie owner Adam Forepaugh (who had an even more elastic relationship with the truth than Barnum) was, as late as 1878, claiming a similar thing about his recently acquired hippo. Forepaugh did eventually get two, one of which died in New York en route to Philadelphia Zoo.[37]

One can trace Bucheet's travels around the United States in regional newspapers (even during the Civil War, when you might expect this kind of activity to have been suspended). He had some exciting adventures, such as an unscheduled swim in the Detroit River in June 1863, when Lee's army had just invaded the north and was pressing into Pennsylvania. Bucheet died at Seeley's Bay in Canada on 7 July 1867, when he would have been eight or at most nine years old. He had lived a short and sad life of relentless exploitation. Bucheet was certainly the first hippopotamus to be seen in the United States and it was not until March 1900 that one was actually born there, in New York's Central Park Zoo. The *Sydney Morning Herald* noted on 6 May of that year that the mother, Mina, was born on the Upper Nile, which makes her a rare late example of a hippo being shipped from Egypt or the Sudan rather than from west or central Africa. Lincoln Park Zoo in Chicago acquired its first hippo even later. She was a Nile hippo called Princess Spearmint

---

37   See Forepaugh's advertisement in the *Los Angeles Herald*, 4 April 1878.

3 A Bloat of Other European Hippos

and came to the zoo as a baby in 1920. Like Obaysch she caused a sensation but, like Obaysch, she also disappointed: when she was first introduced to the waiting public and the flashing bulbs of the press corps she dived into her pool, sank to the bottom, and stayed there.[38]

A female hippopotamus (taken wild in the Congo) was purchased for £850 (plus £300 for freight and the cost of a keeper, Mr Johannson, to accompany her on her long journey) from the German dealer Hagenbeck and arrrived in Adelaide Zoo in South Australia in November 1900 and first went on show on the twenty-fourth of that month. This was a big price: Hagenbeck had originally asked for £900 against an offer of £800. A meeting of the zoo's council authorised a higher bid, which included £200 pounds donated by philanthropic 'colonists', and this was accepted. Much effort and expense to the tune of £269 had been made to build an Egyptian-style hippo enclosure where, as the *South Australian Register* put it, 'she enjoys every comfort so far as it is possible to provide for a creature in captivity.' This was not the first hippopotamus to have visited Australia. The Melbourne *Age* reported that one had been performing with Cooper and Bailey's Circus but died on Christmas Day 1877, although the rival Sells' Circus, an American outfit that visited in November 1891, was reported in the *Sydney Morning Herald* as having the first hippos to have been seen in Australia. The report (which came from the circus itself, so may have had a certain bias) noted that the animal in Cooper and Bailey's show was in a fact a tapir. I am not sure if this is true, as a poster advertising the visit of Cooper and Bailey's to Hobart in April 1877 shows what is clearly a hippo and claims that it had been captured on the White Nile. The Sells' Circus hippos did not perform, but were part of the accompanying menagerie; the same report informs us that the circus had only brought those

---

38 M. Rosenthal, C. Tauber and E. Uhlir, *The Ark in the Park* (Urbana: University of Illinois Press, 2003), pp. 55–56.

animals which were not subject to Australian quarantine regulations. This was also not true, as on the circus's arrival in Sydney all of its horses were placed in quarantine as they showed signs of disease.[39]

Obaysch lived in a Europe that was fascinated by hippos but knew little about them. In every collection baby hippos died soon after birth. Although their bodies added to scientific knowledge of anatomy and physiology, the zoo keepers and naturalists still had very little to go on regarding what was normal hippo behaviour, what was good nutrition, and how best to keep their captives in good health both physically and emotionally. I have argued above that in the case of Obaysch they failed badly, as did all the zookeepers of Europe when it came to hippos. Obaysch's story, and those of the other hippos who dragged out miserable lives in European zoos (and in zoos around the world), offers an exemplary and salutary example in the history of human−animal relationships and the role of captivity in making that a bloody narrative.

While this work was being completed, Barnum's 'Greatest Show on Earth' gave its final performance (on 21 May 2017) as declining audience numbers no longer made for a viable business. It would be comforting to think that this decline was because people had decided that shows involving performing animals had had their day. Unfortunately this is not the case, and it seems that the really damaging decline was due to the removal of performing elephants following successful pressure from animal welfare campaigners. It appears that if the elephants still marched stiffly and sadly round the ring to the crack of whips, people would still go to watch them.

---

39 See M. St Leon, *Circus: The Australian Story* (Melbourne: Melbourne Books, 2011), pp. 118–124.

3 A Bloat of Other European Hippos

If London Zoo can be said to have changed the way people saw animals, this change echoes down to our own day, in the enormous popularity of wildlife documentary films, especially those of Sir David Attenborough. It was Obaysch who was the motor of that change: by looking at Obaysch people learned to see the world differently and to think differently about animals. This influence is still felt today, every time we turn on our television to watch wild animals.

But as we contemplate Obaysch now and attempt to recreate the currents and textures of his life, we learn a different set of lessons and are required to adopt a different way of seeing.

Earlier in this book I suggested that Obaysch was the most important animal of the nineteenth century. Perhaps this was too modest a claim.

Postscript: The Unhappy Hippopotamus

Just as this book was almost finished I came across an attractive little volume in a junk shop. It was the 1965 reprint of Nancy Moore's *The Unhappy Hippopotamus*, which was first published in 1958 when the *New York Times* chose it as one of the best illustrated books of the year.[1] The illustrations are by Edward Leight in a wonderful palette of greys, pink and black that is an unmistakable signature of its period. I grew up in houses that had wallpapers in those very colours. It seemed a serendipitous moment, so, for three dollars, it came home with me.

It turned out that the serendipity was more like synchronicity, as the story of Miss Harriet Hippopotamus was strangely like that of Obaysch. Judging by the details in the illustrations Harriet lives a life of some wealth at some point in the mid Victorian era (she always wears Victorian clothes) and has forgotten how to be a hippopotamus. But although she is surrounded by everything she wants she is unhappy and cannot smile. She meets a mouse who tries to encourage her to try different things but nothing cheers her

---

1 N. Moore, *The Unhappy Hippopotamus* (London: Collins, 1965).

up. Finally the mouse takes her for counselling and the wise old owl advises her that if she wants to be happy she must be a hippo again.

> She then notices that she is standing in mud:
> But why did it feel so good?
> Why? Why, this is why:
> Miss Harriet Hippopotamus
> Was beginning to be a hippo again!
> Once she started being a hippopotamus,
> She couldn't stop being a hippopotamus,
> She didn't WANT to stop.
> She wanted to do EVERYTHING hippopotamuses do.
> She wiggled her toes in the mud.
> She dabbled her feet in the water.
> She threw away her clothes.
> She waded INTO the water.
> Right that second Harriet felt something else –
> A feathery, curling tickling around her mouth!
> Maybe – just possibly – oh, could it really be
> She was going to remember how to smile?
> And then ... and then ... and THEN ...
> Miss Harriet Hippopotamus
> SMILED
> and
> **SMILED**
> and
> **SMILED**

The last illustration shows her in the water naked and with her mouth wide open showing the big hippopotamus teeth which used to be mistaken for tusks.

This is, like Babar's, a story of captivity and release and poses many of the central questions which I have asked about Obaysch in different ways throughout the preceding pages.

Did he know he was a hippopotamus?
Did he remember being a wild hippopotamus?

Postscript: The Unhappy Hippopotamus

I have suggested that Obaysch was the most important animal of the nineteenth century. I have also suggested that he was a fictional character created by Charles Dickens. Perhaps I could also suggest that although his fame depended on his being a hippo, he was not a hippo at all but rather a hippo-shaped bundle of sensation and instinctive aggression.

I would ask, was he happy? Happiness – perhaps joy is a better word – seems, from observation, to be something which animals experience and which we can attribute to them without being too mawkish or projecting our anthropomorphic fantasies onto them. But I'm afraid I think I do know the answer to that one and I fear that, like Harriet, Obaysch – if he was a hippo – was an unhappy hippopotamus.

Obaysch and His Bloat

Obaysch, born Sudan 1849, died London 11 March 1878.

Adhela, born Sudan 1853, died London 16 December 1883.

Unnamed male, born London 21 February 1871, died London 23 February 1871.

Unnamed female (Umzimvooboo?), born London 6 January 1872, died London 10 January 1872.

Guy Fawkes, born London 5 November 1872, died London 20 March 1908.

A Note on Sources

Most of the original material for this work comes from contemporary newspapers and magazines. Quotations from these periodicals are not generally footnoted but are referenced within the text. They are not listed in the bibliography, which lists only books, book chapters and scholarly articles. Translations from French and German are my own.

Newspaper and magazine articles were sourced mainly from the following databases:

American: chroniclingamerica.loc.gov, although many titles have their own archives and there are also other sources such as ucsd.libguides.com and newspaperarchive.com
Australian and New Zealand: trove.nla.gov.au
Belgian: opac.kbr.be
British: gale.cengage.co.uk
Dutch: delpher.nl
French: gallica.bnf.fr
German: eudocs.lib.byu.edu
Other European newspapers: theeuropeanlibrary.org

The Hathi Trust Digital Library (https://m.hathitrust.org) is also full of valuable material, especially for the publications of the Zoological Society of London during the relevant period and beyond.

Bibliography

Allen, E.A. *The Victorian Fern Craze: A History of Pteridomania* (London: Hutchinson, 1969).
Allin, M. *Zarafa: A Giraffe's True Story from Deep in Africa to the Heart of Paris* (London: Headline, 1998).
Amato, S. *Beastly Possessions: Animals in Victorian Consumer Culture* (Toronto: University of Toronto Press, 2015).
Ames, E. *Carl Hagenbeck's Empire of Entertainments* (Seattle: University of Washington Press, 2009).
Anderson, N., and P. Steeves. *Political Animals* (Chicago: Northwestern University Press, forthcoming).
Anderson, V.D. *Creatures of Empire: How Domestic Animals Transformed Early America* (Oxford: Oxford University Press, 2006).
Auerbach, J.A. *The Great Exhibition: A Nation on Display* (New Haven, CT: Yale University Press, 1999).
Bankoff, G., and S. Swart. *Breeds of Empire: The 'Invention' of the Horse in South-East Asia and Southern Africa* (Copenhagen: NIAS Press, 2007).
Baratay, E., and E. Hardouin-Fugier. *Zoo: A History of Zoological Gardens in the West* (London: Reaktion, 2003).
Barber, L. *The Heyday of Natural History, 1820–1870* (London: Jonathan Cape, 1980).
Barnaby, D. *The Elephant Who Walked to Manchester* (Plymouth: Basset Publications, 1988).
Barrington-Johnson, J. *Zoo: The Story of London Zoo* (London: Robert Hale, 2005).

Barrow, J. *Characteristic Sketches of Animals Principally from the Zoological Gardens Regent's Park* (London: Moon, Boys & Glaves, 1832).
Barry, A. *The Life and Works of Sir Charles Barry* (London: James Murray, 1867).
Bartlett, A.D. *Wild Animals in Captivity* (London: Chapman & Hall, 1899).
Bartlett, D.W. *What I Saw in London* (New York: C.M. Saxton, Barker & Co., 1852).
Bedini, S.A. *The Pope's Elephant* (Manchester: Carcanet, 1997).
Belozerskaya, M. *The Medici Giraffe* (New York: Little, Brown & Co., 2006).
Benbow, M.S.P. 'Death and Dying at the Zoo', *Journal of Popular Culture* 37 (2004), pp. 379–398.
Bingley, W. *Animal Biography, or, Authentic Anecdotes of the Lives, Manners and Economy of the Animal Creation* (London: Richard Phillips, 1803).
Blanchard, P., N. Bancel, G. Boëtsch, E. Deroo, S. Lemaire and C. Forsdick (eds). *Human Zoos* (Liverpool: Liverpool University Press, 2008).
Blaszkiewitz, B. *Knautschke, Knut & Co.: Die Lieblingstiere der Berliner Aus Tierpark und Zoo* (Berlin: Lehmanns, 2008) [Knautschke, Knut & Co.: The Best-Loved Animals of the Berlin Animal Park and Zoo].
Blunt, W. *The Ark in the Park: The Zoo in the Nineteenth Century* (London: Hamish Hamilton, 1976).
Bompas, G. *The Life of Frank Buckland* (London: Smith, Elder & Co., 1886).
Boomgard, P. *Frontiers of Fear* (New Haven, CT: Yale University Press, 2001).
Bostock, F. *The Training of Wild Animals* (New York: The Century Co., 1904).
Boswell, J. *The Life of Samuel Johnson*. Edited by R. Ingpen. (London: George Bayntun, 1925), 2 volumes.
Boyde, M. *Captured* (London: Palgrave Macmillan, 2014).
Bridger, B. *Buffalo Bill and Sitting Bull: Inventing the Wild West* (Austin: University of Texas Press, 2002).
Bruce, G. *Through the Lion Gate: A History of Berlin Zoo* (Oxford: Oxford University Press, 2017).
Buckland, F.T. *Curiosities of Natural History*, 3rd Series (London: Richard Bentley, 1868) .
Burchell, W.J. *Travels in the Interior of Southern Africa* (London: Longman, Hurst, Rees, Orme & Brown, 1822).
Burkhardt Jr, R.W. 'Constructing the Zoo: Science, Society and Animal Nature at the Paris Menagerie 1794–1838', in M. Henniger-Voss (ed.), *Animals in Human History* (Rochester: University of Rochester Press, 2002), pp. 231–257.
Cadbury, D. *The Dinosaur Hunters* (London: Fourth Estate, 2001).
Chalmers-Mitchell, P. *The Zoological Society of London: Centenary History* (London: Zoological Society of London, 1929).

Bibliography

Chambers, P. *Jumbo* (London: André Deutsch, 2007).
Chitty, N., L. Ji, G. Rawnsley and C. Hayden (eds). *The Routledge Handbook of Soft Power* (London: Routledge, 2017).
Church, S.K. 'The Giraffe of Bengal: A Medieval Encounter in Ming China', *The Medieval History Journal* 7 (2004), pp. 2–39.
Colley, A.C. *Wild Animal Skins in Victorian Britain* (Farnham, UK: Ashgate Publishing Ltd., 2014).
Cooper, R. *Roman Antiquities in Renaissance France* (Farnham, UK: Ashgate Publishing Ltd., 2013).
Cowie, H. *Exhibiting Exotic Animals in Nineteenth-Century Britain* (London: Palgrave Macmillan, 2014).
Crosby, A. *Ecological Imperialism* (Cambridge: Cambridge University Press, 2004).
Crossley, C. *Consumable Metaphors* (Oxford: Peter Lang, 2005).
Dampier, W. *Voyages and Descriptions* (London: James and John Knapton, 1699).
Dance, S.P. *Shell Collecting* (London: Faber & Faber, 1966).
Daniel, J.C., and B. Singh. *Natural History and the Indian Army* (Mumbai: Oxford University Press for Bombay Natural History Society, 2009).
Davis, R.H. *The Congo and Coast of Africa* (New York: Charles Scribner's Sons, 1907).
de Courcy, C. 'Evolution of a Zoo: A History of the Melbourne Zoological Gardens, 1857–1900' (MA thesis, University of Melbourne, 1991).
de Courcy, C. *The Zoo Story: The Animals, the History, the People* (Melbourne: Penguin, 1995).
de Courcy, C. *Dublin Zoo* (Wilton: The Collins Press, 2009).
Deiss, W., and R. Hoage (eds). *New Worlds, New Animals* (Baltimore: Johns Hopkins University Press, 1996).
Desmond, A. 'The Making of Institutional Zoology in London 1822–1836', *History of Science* 23 (1985), pp. 133–185.
Donald, D. *Picturing Animals in Britain* (New Haven, CT: Yale University Press, 2007).
Drew, W.A. *Glimpses and Gatherings* (Augusta, ME: Homan & Markey, 1852).
Edwards, H. Sutherland. *Old and New Paris* (London: Cassell, 1892–1994).
Etringham, S.K. *The Hippos* (London: Bloomsbury, 1999).
Ferguson, M. *Animal Advocacy and Englishwomen 1780–1900* (Ann Arbor: University of Michigan Press, 1998).
Fisher, C. (ed.). *A Passion for Natural History* (Liverpool: National Museums & Galleries on Merseyside, 2002).
Flack, A.J.P. '"The Illustrious Stranger": Hippomania and the Nature of the Exotic', *Anthrozoös* 26 (2013), pp. 43–59.
Forbes, C. *Australia on Horseback* (Sydney: Macmillan, 2014).

Franks, A.E. 'The Pursuit of Hippo-ness: Hippopotamus and Human' (MFA thesis, Montana State University, 2014).
Gail, M. *Persia and the Victorians* (London: Routledge, 2013).
Gallop, A. *Buffalo Bill's British Wild West* (Stroud, Glos.: The History Press, 2009).
Gates, B.T. *Kindred Nature* (Chicago: University of Chicago Press, 1998).
Germond, P. *An Egyptian Bestiary* (London: Thames & Hudson, 2001).
Girling, R. *The Man Who Ate the Zoo* (London: Random House, 2016).
Glendinning, V. *Raffles and the Golden Opportunity* (London: Profile, 2012).
Gorrie, D. *Geordie Purdie in London* (London: Houlston & Son, 1875).
Graham, F. *Visits to the Zoological Gardens* (London: George Routledge, 1853).
Grigson, C. *Menagerie* (Oxford: Oxford University Press, 2016).
Hahn, D. *The Tower Menagerie* (London: Pocket Books, 2003).
Halls, J.J. *The Life and Correspondence of Henry Salt* (London: Richard Bentley, 1834).
Hancock, D. *A Different Nature* (Berkeley: University of California Press, 2001).
Haney, J. *Haney's Art of Training Animals* (New York: J. Haney & Co., 1869).
Harding, L. *Elephant Story* (Jefferson, NC: MacFarland and Company, 2000).
Henniger-Voss, M. *Animals in Human History* (Rochester, NY: University of Rochester Press, 2002).
Hill, C.V. 'Colonial Gardens and the Validation of Empire in Imperial India', *Journal of South Asian Studies* 1 (2013), pp. 139–145.
Hoage, R.J., and W.A. Deiss (eds). *New Worlds, New Animals* (Baltimore; Johns Hopkins University Press, 1996).
Homans, M., and A. Munich. *Remaking Queen Victoria* (Cambridge: Cambridge University Press, 1997).
Jackson, C.E. *Menageries in Britain 1100–2000* (London: The Ray Society, 2014).
Jarofke, D. *Das Flusspferd Knautschke, unser Freundlicher Nachbar* (Berlin: Schuling, 2012) [The Hippopotamus Knautschke, Our Friendly Neighbour].
Jeal, T. *Livingstone* (London: Pimlico, 1993).
Jeal, T. *Stanley* (London: Faber & Faber, 2011).
Jeal, T. *Explorers of the Nile* (London: Faber & Faber, 2011).
Johnson, C. *Australia's Mammal Extinctions* (Melbourne: Cambridge University Press, 2006).
Jones, R.W. 'The Sight of Creatures Strange to Our Clime': London Zoo and the Consumption of the Exotic', *Journal of Victorian Culture* 2 (1997), pp. 1–26.
Kaeppler, A.L. *Holophusicon* (Altenstadt, HE: ZFK Publishers, 2012).
Kenny, R. *The Lamb Enters the Dreaming* (Melbourne: Scribe, 2007).
Kete, K. *The Beast in the Boudoir* (Berkeley: University of California Press, 1994) .
Kete, K. (ed.). *A Cultural History of Animals in the Age of Empire* (Oxford: Berg, 2007).

Bibliography

Kingsley, M. *Travels in West Africa* (London: Macmillan & Co. Ltd.: 1897).'
Kisling, V.N. *Zoo and Aquarium History* (Boca Raton, FL: CRC Press, 2001).
Kisling, V.N. 'Colonial Menageries and the Exchange of Exotic Faunas', *Archives of Natural History* 25 (2000), pp. 303-320.
Lever, C. *They Dined on Eland* (London: Quiller Press, 1992).
Lloyd, J.B. *African Animals in Renaissance Art* (Oxford: The Clarendon Press, 1971).
Lobo, J. *A Voyage to Abyssinia* (London: Elliot & Gay, 1789).
Loisel, G. *Histoire des Ménageries de l'Antiquité à Nos Jours*, 3 volumes (Paris: O. Doin et Fils, 1912).
Low, D.A. *Buganda in Modern History* (Berkeley: University of California Press, 1971).
Luckhurst, R., and J. McDonagh (eds). *Transactions and Encounters: Science and Culture in the Nineteenth Century* (Manchester: Manchester University Press, 2002).
Mabille, G., and J. Pieragnoli. *La Ménagerie de Versailles* (Arles, PACA: Éditions Honoré Clair, 2010).
Mackenzie, C. and S. Posthumus (eds). *French Thinking about Animals*. (East Lancing: Michigan State University Press, 2015).
MacKenzie, J.M. *The Empire of Nature* (Manchester: Manchester University Press, 1988).
MacKenzie, J. M. (ed.). *Imperialism and the Natural World* (Manchester: Manchester University Press, 1990).
McDowall, R. *Gamekeepers for the Nation* (Christchurch: Canterbury University Press, 1994).
Maddox, B. *Reading the Rocks* (London: Bloomsbury, 2017).
Malamud, R. *Reading Zoos* (New York: New York University Press, 1998).
Manley, D. 'A Traveller from Egypt', *Bulletin of the Association for the Study of Travel in Egypt & the Near East: Notes & Queries* 21 (2004), pp. 21-22.
Marino, L., S.O. Lilienfeld, R. Malamud, N. Nobis and R, Broglio, 'Do Zoos and Aquariums Promote Attitude Change in Visitors?', *Society & Animals* 18 (2010), pp. 126-138.
Maxwell, H. *The Honourable Sir Charles Murray KCB: A Memoir* (Edinburgh: W. Blackwood, 1898).
Mehos, D. *Science and Culture for Members Only* (Amsterdam: Amsterdam University Press, 2005).
Middlemiss, J.L. *A Zoo on Wheels* (Burton-on-Trent, Staffs.: Dalebrook Publications, 1987).
Mikhail, A. *The Animal in Ottoman Egypt* (Oxford: Oxford University Press, 2014).

Montagnana-Wallace, N. *150 Years Melbourne Zoo* (Thornbury: Bounce Books, 2012).
Mooallem, J. 'American Hippopotamus', *Atavist Magazine* 32 (2013), https://bit.ly/2iVT7lo.
Moore, N. *The Unhappy Hippopotamus* (London: Collins, 1965).
Morris, P.A. *Van Ingen and Van Ingen* (Ascot: MPM Publishing, 2006).
Morse, D.D., and M.A. Danahay. *Victorian Animal Dreams* (Aldershot, Hants.: Ashgate Publishing Ltd., 2007).
Morton, M. (ed.). *Oudry's Painted Menagerie* (Los Angeles: Getty Publications, 2007).
Mullan, B., and G. Marvin. *Zoo Culture* (Urbana: University of Illinois Press, 1999).
Murray, N. 'Lives of the Zoo: Charismatic Animals in the Social World of the Zoological Gardens of London, 1850–1897' (PhD thesis, Indiana University, 2004).
Nicholls, S. *Paradise Found: Nature in America at the Time of Discovery* (Chicago: University of Chicago Press, 2011).
Noble, J. *Around the Coast with Buffalo Bill* (Beverley, Yorks.: Hutton Press Ltd, 1999).
Northrop, H.D. *Earth, Sky and Sea* (San Francisco: The J. Dewing Co., 1887).
Plumb, C. *The Georgian Menagerie* (London: I.B. Tauris, 2015).
Quick, T. 'Interpretations of London's Zoological Hippopotami' (MSc thesis, London Centre for the History of Science, 2007).
Parsonson, I. *The Australian Ark* (Collingwood, Vic.: CSIRO, 2000).
Picker, J.M. *Victorian Soundscapes* (Oxford: Oxford University Press, 2003).
Polack, E. *Kako Le Terrible* (Paris: La Joie de Lire, 2013).
Reynolds, R. 'America's First Hippo', *Association of Zoos and Aquariums Regional Conference Proceedings* (AZA, 1996), pp. 346–351.
Ridley, G. *Clara's Grand Tour* (London: Atlantic Books, 2004).
Ritvo, H. *The Animal Estate: The English and Other Creatures in the Victorian Age* (Cambridge, MA: Harvard University Press, 1987).
Ritvo, H. *The Platypus and the Mermaid* (Cambridge, MA: Harvard University Press, 1997).
Robbins, L.E. *Elephant Slaves and Pampered Pets* (Baltimore: Johns Hopkins University Press, 2002).
Rockwell, D. *Giving Voice to Bear* (Lanham, MD: Roberts Rinehart Publishers, rev. ed. 2003).
Rolls, E. *They All Ran Wild* (London: Angus & Robertson, 1984)
Rookmaker, K. 'The Royal Menagerie of the King of Oudh', *Back When, and Then* 2:1 (August 1997), p. 10.

Bibliography

Root, N. J. 'Victorian England's Hippomania', *Natural History* 103 (1993), pp. 34–39.
Rosenthal, M., C. Tauber and E. Uhlir. *The Ark in the Park* (Urbana: University of Illinois Press, 2003).
Rothfels, N. *Savages and Beasts: The Birth of the Modern Zoo* (Baltimore and London: Johns Hopkins University Press, 2002).
Rothfels, N. 'Catching Animals', in M. Henniger-Voss (ed.), *Animals in Human History* (Rochester: University of Rochester Press, 2002), pp. 182–228.
Rothfels, N. 'How the Caged Bird Sings: Animals and Entertainment', in K. Kete (ed.), *A Cultural History of Animals in the Age of Empire* (Oxford: Berg, 2011), pp. 95–112.
Rothschild, M. *Walter Rothschild* (London: Natural History Museum, 2008).
Sahlins, P. *1668: The Year of the Animal in France* (New York: Zone Books, 2017).
Scherren, H. *The Zoological Society: A Sketch of its Foundation and Developments* (London: Cassell, 1905).
Schiebinger, L. *Plants and Empire* (Cambridge, MA: Harvard University Press, 2004).
Schomburgk, H. *Wild und Wilde im Herzen Afrikas* (Berlin: Deutsche Buch-Gemeinschaft, 1925).
Scrivens, K., and S. Smith. *Manders Shows and Menageries* (Newcastle-under-Lyme, Staffs.: The Fairground Society, no date).
Shipman, P. *To the Heart of the Nile* (New York: Morrow, 2004).
Simons, J. *Rossetti's Wombat* (London: Middlesex University Press, 2008).
Simons, J. *The Tiger that Swallowed the Boy* (Faringdon, Oxon.: Libri, 2012).
Simons, J. 'The Scramble for Elephants: Exotic Animals and the Imperial Economy', in M. Boyde, *Captured* (London: Palgrave Macmillan, 2014), pp. 24–42.
Simons, J. 'The Soft Power of Elephants', in N. Chitty, L. Ji, G. Rawnsley and C. Hayden (eds), *The Routledge Handbook of Soft Power* (London: Routledge, 2017), pp. 177–184.
Smith, A. *Illustrations of the Zoology of Southern Africa* (London: Smith, Elder & Co, 1838–1849).
Sparrman, A. *A Voyage to the Cape of Good Hope* (London: G.G.J. & J. Robinson, 1785).
St Leon, M. *Circus, the Australian Story* (Melbourne: Melbourne Books, 2011).
Stott, R. 'Darwin's Barnacles', in R. Luckhurst and J. McDonagh (eds), *Transactions and Encounters: Science and Culture in the Nineteenth Century* (Manchester: Manchester University Press, 2002), pp. 151–181.
Stott, R. *Theatres of Glass* (London: Short Books, 2003).
Sutherland, J. *Jumbo* (London: Aurum Press Ltd, 2014).

Tague, I.H. *Animal Companions* (University Park: Pennsylvania State University Press, 2015).
Tait, P. *Fighting Nature* (Sydney: Sydney University Press, 2016).
Takashi, I. *London Zoo and the Victorians 1828–1859* (London: Royal Historical Society, 2014).
Thomas, K. *Man and the Natural World* (Oxford: Oxford University Press, 1983).
Thompson, Z. *Journal of a Trip to London, Paris and the Great Exhibition* (Burlington: Nichols & Warren, 1852).
Thornbury, W. *Old and New London*, 6 volumes, 5 (London: Cassell & Co., 1887–1893).
Toman, J. *Kilvert's World of Wonders: Growing Up in Mid-Victorian England* (London: Lutterworth Press, 2014).
Toynbee, J.M.C. *Animals in Roman Life and Art* (Baltimore: Johns Hopkins University Press, 1973).
Trafton, M. *Rambles in Europe* (Boston: Charles H. Pierce & Co., 1852).
Trow, M.J. *The Adventures of Sir Samuel White Baker, Victorian Hero* (London: Pen & Sword Books, 2010).
Twain, M. *Mark Twain's Letters*. Edited by E.M. Branch et al. Berkeley: University of California Press, 1997.
Velten, H. *Beastly London* (London: Reaktion, 2013).
Velvin, E. *Wild Animal Celebrities* (New York: Moffatt, Yard & Co., 1907).
Verney, P. *Homo Tyrannicus* (London: Mills & Boon, 1979).
Vevers, G. *London's Zoo* (London: Bodley Head, 1976).
von Uffenbach, Z. *Merkwurdige Reisen durch Niedersachsen, Holland und Engeland* (Noteworthy travels through Lower Saxony, Holland and England) (Frankfurt and Leipzig, 1754).
Walker, B.L. *The Lost Wolves of Japan* (Seattle: University of Washington Press, 2005).
Walker, S., A. Pal, B. Rathanasabapathy and R. Manikam. 'Indian Zoological and Botanical Gardens – Historical Perspective and a Way Forward', *BGjournal* 1 (2004), no page nos.
Warden, J.S. 'Lord Byron and the Hippopotamus', *Notes and Queries* 12 (1855).
Watkin, F. *The Year of the Wombat* (London: Gollancz, 1974).
Wilkinson, A. *The Passion for Pelargoniums* (Stroud, Surrey: Sutton Publishing, 2007).
Williams, E. *Hippopotamus* (London: Reaktion, 2017).
Wilson, A.N. *The Victorians* (London: Hutchinson, 2007).
Wynter, A. *Curiosities of Civilisation* (London: R. Hendricks, 1860).
Yallop, J. *Magpies, Squirrels and Thieves* (London: Atlantic Books, 2011).

Index

Abbas I, Pasha 36–48, 78, 80
Abbott, Henry 47
acclimatisation societies 29, 90, 169
Adhela (hippopotamus)
  aggressive nature of 81, 91, 98, 101,
    105, 152
  capture of 79
  death of 111–112
  offspring 30, 148, 155, 176, 189
  transportation to London 79
Albert, Prince 8, 23
Alhambra Palace 177–179
Alma Tadema, Lawrence 198
Amato, Sarah 7
animal handlers 38, 51, 55, 75
animal performance 47, 66–71, 79,
  178–179, 183, 201, 202
animal training 37, 155, 178
ant-eaters 30, 153, 161
Anthony (hippopotamus) 108
anthropomorphism 8, 140, 188, 207
Antoine (hippopotamus) 196
aquariums 162, 179
audience behaviour 68, 84, 152

Babar the Elephant 15, 206
Baden-Powell, Robert 89
Baker, Samuel and Lady 149
Bankoff, Greg 14
Baptiste (hippopotamus) 192, 196
Barnum, P.T. 17, 86, 106, 179, 200, 202
  Greatest Show on Earth 179, 202
Barrow, John 142
Bartlett, Abraham 15, 81, 83, 88–114,
  145, 176, 187, 189
Bartlett, Edward 82
Barry, Charles 19
Belon, Pierre 146
Berners, Lord 3
Bichette (hippopotamus) 27, 187,
  189–197
Blake, William 141
Blunt, Wilfred 6
Boomgaard, Peter 14
Borbón, Don Juan Carlos María Isidro
  de 83
Braner, Monsieur 191, 196
Britain 1, 14–18, 23, 27, 35–38, 135
  in the Roman era 2, 16, 22

221

British Empire 12, 13–14, 27, 67
  in Africa 12, 13–14, 37, 151
  in Australasia 11, 12, 27, 54
  in India 12–13, 27, 54
Bruce, Frederick William 78
Brunhoff, Jean de *see Babar the Elephant*
Brutus (hippopotamus) 174
Bucheet (hippopotamus) 85, 87, 179, 200
Buckland, Frank 16, 29, 50, 84–106, 176, 179, 186, 190
Buffalo Bill 10
Buffon, Comte de 142, 144
Burchell, William John 142
Burton, Sir Richard 149
Byron, Lord 21–22

Cannani, Hamet 15, 46–79, 138, 147, 151, 154, 159, 176
cartoons 8, 67–69, 77, 96, 110, 152, 160, 161
children's books 156, 157, 158, 198, 205; *see also Babar the Elephant*
Chunee (elephant) 93
circus animals 87, 111, 179, 201; *see also* Barnum, P.T., Cooper and Bailey's Circus
class 21, 23, 25, 27–31, 136
Coco (hippopotamus) 27, 184–186, 194–196
collecting animals 18, 24, 54, 72, 75, 86, 162, 183
Cooper and Bailey's Circus 201
colonialism 6, 10–16, 151, 154, 172, 197; *see also* British Empire
Commodus, Emperor 2
Conrad, Joseph 19
Cosimo III 18
Cowie, Helen 7

Crisp, Edward 182
Crosby, Alfred 11
Cross, William 180

Daniell, Samuel 20, 118
Dickens, Charles 9, 24, 63, 72, 86, 113, 133–140, 151, 207
  Household Words 135–140
Dorothea (hippopotamus) 199

Edwards, Henry Sutherland 191
Egypt 3, 5, 13, 19, 34–49, 75–78, 185
elephants 4, 15, 31, 68–72, 93, 161
empire 7–16, 65, 189; *see also* British Empire, Ottoman Empire
ethnographic display 10, 19

Fantasia (film) 175
feeding display 71, 139
Fillis, Frank E. 11, 12
Firmus 3
Flack, Andrew 6
Forepaugh, Adam 200
France 18, 27, 64, 83, 183–185, 193–195
  expansion into Africa 13, 75

giraffes 4, 70–75, 184
Gordian I 3
Gordian III 2
Gordon-Cumming, Roualeyn George 56
gorillas 152
Gorrie, Daniel 140
Gould, John 20, 72, 73, 184
Great Britain *see* Britain
Great Exhibition of 1851 20, 23, 73, 76, 90
Grigson, Caroline 7
Gustavito (hippopotamus) 174

Index

Guy Fawkes (hippopotamus) 23, 100–107, 112–115, 143, 145, 152, 157–159, 169, 176, 189, 209

Hagenbeck, Carl 11, 199, 201
Haijab, Jabar Abou 75–76
Halliwell-Philips, Henrietta 156
Hamet *see* Cannani, Hamet
Heber, Bishop Reginald 149–150
Hermann (hippopotamus) 199
'hippomania' 6, 115, 162–163, 169, 194, 195
hippopotamuses
 aggressive nature of 106, 113, 187, 191, 198
 as Behemoth 141–141, 154
 lifespan of 107, 111
 size of 57, 71, 109
 Victorian knowledge of 141, 175, 177, 202
Horne, Richard Henry 137–138
horses 37, 202
Household Words see Dickens, Charles: *Household Words*
Howitt, William Samuel 20
Huberta (hippopotamus) 173
hummingbirds 31, 72
Hunt, Henry 82, 88, 91
hunting 37–45, 146, 154, 181

Ibrahim 36, 48
Illustrated London News 26, 38, 68, 80, 83, 99, 103, 169
imperialism *see* colonialism, empire
India *see* British Empire: in India
Ito, Takashi 6

Jabar *see* Haijab, Jabar Abou
Jackson, Christine 7
Japan 12, 14

Jardin d'Acclimatation 30, 196, 196
Jardin des Plantes *see* zoological gardens: Jardin des Plantes, Paris
Jupiter (hippopotamus) 115

Kako (hippopotamus) 196–198
Kenny, Robert 14
Kilvert, Emily 67
Kilvert, Reverend Francis 68
Kingsley, Mary 90
Knautschke (hippopotamus) 172–173

Lancelle, Jean-Baptiste 197
Landy, Monsieur 197
Lawrence, Elizabeth 157, 158
Leclerc, G.L. *see* Buffon, Comte de
leisure market 23, 28, 31
Lever, Ashton
 Holophusikon, the 19
lions 51, 65, 152, 160, 165
Livingstone, David 154, 156
Lobo, Father Jeronimo 142
Loisel, G. 195
London Zoo *see* Regent's Park Zoological Gardens
Lotus (hippopotamus) 111
Louisette (hippopotamus) 196

Malamud, Randy 7
media 9, 91, 107, 135, 139
menageries *see also* zoological gardens
 Chantilly 183
 Exeter 'Change 21, 93
 Giardino del Boboli 18
 Tiergarten Schönbrunn 19, 172, 183
 Versailles 183
Merwan, Mohammed Abou 62, 75–79
Mikhail, Alan 14
Milne-Edwards, Alphonse 196

Mitchell, David 15, 29–31, 42, 62, 64, 161
monkeys 24, 31
Moresby, Captain John 49
Morrison, Arthur 113–114
Morton, Samuel G. 35
Murray, Charles 'Hippopotamus' 15, 36, 45–48, 73, 76, 78, 143, 145, 151–151
Murray, Narisara 6, 151

Napoleon, Louis (Napoleon III) 34, 75, 186
'Naturalist, The' (columnist) 62–63
Nepal 50, 54
Newsboy (hippopotamus) 174
Nile River 5, 33, 37, 58, 69, 86
　White Nile 41, 49, 71, 101, 109
Nina (hippopotamus) 175

Obaysch
　agressive nature of 90, 92, 109, 187
　before his capture 37–41
　capture of 33, 42–44, 146
　death of 108–112
　depictions of 140, 151, 160
　early writing about 134–140
　journey to London 45–55
　nocturnal habits of 59, 66, 74
　size and weight 106
　zoo escape 88–89
ostriches 3, 38, 55
Ottoman Empire 13–14, 35, 149
Owen, Richard 57–60, 68, 156, 159

Patrick (wombat) xiii
'Peter Possum' (columnist) 92
Petherick, John 33, 85, 178–179
People's Spring, The 34
Plumb, Christopher 7

Prescot, Michael 106
press *see* media
Princess Spearmint (hippopotamus) 200
Punch 30, 48, 56–59, 69, 77, 83, 90, 110–111, 151–153, 159–161, 165–168, 184

Quick, G.C. 200
Quick, Tom 6

Ranagee, Jung Koorman Bahadoor 50
Rankin, F. Harrison 150
Regent's Park Zoological Gardens 6, 21, 24, 26, 33, 59, 68, 99, 111, 143, 179, 183, 185, 198
rhinoceroses 4, 49, 59, 99, 157, 180
Ripon, S.S. 49–54, 75, 78, 145, 161, 167
Ritvo, Harriet 11
Rockwell, David 14
Root, Nina 6, 162
Rossetti, Dante Gabriel xi, 8, 29
Rothsfels, Nigel 7
Rubens, Peter Paul 4
Rüppell, Eduard 43, 44
Russia 35, 166

Saint-Hilaire, Geoffroy 75
Salama (keeper) 86–87, 179
Salt, Henry 145
Scott, Matthew 88–89
sea lions 24, 34
Sells' Circus 201
Shepheard, Samuel 46, 168
Simons, John
　Rossetti's Wombat xi
　The Tiger that Swallowed the Boy xi
slave trade 14, 149, 153, 181, 189
Smith, Andrew 143
snake charmers 75, 151
snakes 51, 75

Index

souvenirs 73, 165
Sparrman, Anders 142, 145
speciesism 10, 13
St Mars, Louis
   Hippopotamus Polka 60, 163
'star' animals 6, 30–31, 68, 71, 76–78, 137, 145, 156, 161
Sudan 14, 44, 77, 178, 200
Suez Canal 27, 54, 179, 181
Susie (hippopotamus) 174
Swart, Sandra 14

Tait, Peta 7
Tanja (hippopotamus) 174
tapirs 22, 144, 201
taxidermy 5, 19, 97, 169, 173; *see also* Gould, John
Taweret, the hippo god/goddess 3
theatre 31, 136, 164, 177
Thomas, Keith 11
Thompson, Zadock 65, 78
Thomson, Arthur 106
Thornbury, William 139
Thunberg, Carl Peter 145
Top (wombat) xiv, 8
trade in wild animals 27, 54, 180–181
Trafton, Mark 78
transportation of wild animals 33, 43–54, 79, 137, 144, 179, 180, 193, 196
travelling menageries 11, 25, 179, 200
Trelawny, Edward 85

United Kingdom *see* Britain

Victoria, Queen 2, 15, 50, 61, 68, 72, 75, 151
Victorian era 2, 6–9, 23–28, 48, 70, 76, 133–137, 141, 148–155, 161, 171

Walker, Brett 14
Warden, J.S. 21–22
Wild West Show 11
Wolf, Joseph 143
wombats xiii, 29, 74
Wright-Bruce, Frederick William *see* Bruce, Frederick William
Wynter, Andrew 82–83, 91

York, Frederick 95

Zerenghi, Federico 5
zoological gardens *see also* menageries
   Adelaide Zoo 29, 71, 174
   admission prices 19, 21, 24, 28, 72
   Belle Vue Zoo, Manchester 24
   Berlin Zoo 172–175
   Central Park Zoo, New York 200
   Antwerp Zoo 115, 196, 199
   Amsterdam Zoo 4, 16, 27, 100, 107, 115, 174, 176, 182, 190, 199
   Jardin des Plantes, Paris 6, 26–27, 65, 80, 167, 183–197
   Lincoln Park Zoo, Chicago 200
   Melbourne Zoo 11, 29, 175
   proliferation of 9, 12, 23
   Regent's Park Zoological Gardens, London *see* Regent's Park Zoological Gardens
   San Salvador National Zoological Park, El Salvador 174
   science and education in 6, 13, 23, 28, 67, 157
   zookeepers 29, 88, 98, 107, 176, 184, 187, 192, 202
   killed by hippopotamuses 191, 196, 197
Zoological Society of London 8, 16, 17, 21, 25–28, 32, 36, 45–49, 59, 66,

Obaysch

71–109, 112, 137, 144, 148, 168, 176, 179, 189, 193

www.ingramcontent.com/pod-product-compliance
Lightning Source LLC
Chambersburg PA
CBHW050522170426
43201CB00013B/2051